RAND NATIONAL DEFENSE RESEARCH INSTITUTE

The Federal Civil Service Workforce

Assessing the Effects on Retention of
Pay Freezes, Unpaid Furloughs, and Other
Federal-Employee Compensation Changes
in the Department of Defense

Beth J. Asch, Michael G. Mattock, James Hosek

Prepared for the Office of the Secretary of Defense

For more information on this publication, visit www.rand.org/t/rr514

Library of Congress Cataloging-in-Publication Data
ISBN: 978-0-8330-8685-3

Published by the RAND Corporation, Santa Monica, Calif.
© Copyright 2014 RAND Corporation
RAND® is a registered trademark.

Cover Image: *Николай Григорьев/Fotolia, jameschipper/Fotolia*

Support RAND
Make a tax-deductible charitable contribution at
www.rand.org/giving/contribute

www.rand.org

Preface

Federal-employee pay freezes between 2011 and 2013, together with unpaid furloughs brought about by sequestration and other federal-budget-related activities, have raised concerns about how these changes in compensation will affect the ability of the federal civil service to attract and retain personnel. These concerns are part of a broader set of questions about setting the adequate level and structure of federal compensation and how changes in federal-employee compensation will affect the current and future workforce size and experience mix.

To understand the effects of changes in the level and structure of compensation on civil service retention, a model is needed that is estimated with data on the individual careers of civil service employees and that permits analyses of the retention effects of compensation changes. Few studies have estimated the effects of civil service compensation, and the studies that do exist use a methodology with distinct drawbacks. The study reported here begins to fill the gap by using a methodology that addresses these drawbacks, analyzing retention dynamics that result from changes in financial incentives to serve in the federal civil service, using data on defense General Schedule employees. The study is an initial foray into developing a capability for modeling the responsiveness of federal civil service retention to changes in compensation.

This report demonstrates the capability developed in the study by considering the effects on retention of recent highly visible policy changes—the federal pay freeze and unpaid furloughs. The model development and policy analysis should be of interest to the policy community concerned with the effectiveness of federal compensation, as well as the research community concerned with human resource and personnel issues.

This research was conducted within the Forces and Resources Policy Center of the RAND National Defense Research Institute, a federally funded research and development center sponsored by the Office of the Secretary of Defense, the Joint Staff, the Unified Combatant Commands, the Navy, the Marine Corps, the defense agencies, and the defense Intelligence Community.

For more information on the Forces and Resources Policy Center, see http://www.rand.org/nsrd/ndri/centers/frp.html or contact the director (contact information is provided on the web page).

Contents

vi   Federal Civil Service Retention

Figures

Tables

Summary

The federal civil service workforce in the United States is large and, in the context of the Department of Defense (DoD), an important contributor to military readiness. It is therefore incumbent on planners and policymakers to ensure that this workforce is managed effectively, fairly, and at a minimum cost to the taxpayer. Retention is a critical part of workforce management. Federal agencies must ensure that they recruit and retain personnel with the right skills, capabilities, and experience levels to meet their workforce requirements. Understanding retention is particularly important because of the 2011–2013 pay freeze, the unpaid furloughs in 2013, a wave of retirements of the baby-boom generation, and recent proposals by lawmakers to cut federal compensation, including changes to the federal retirement system.

Planners and policymakers need to be able to assess how compensation policy, including the recent pay freezes and unpaid furlough, affects federal civil service retention. The research summarized in this report begins to meet this need. It extends a modeling capability used successfully in analyses of retention in both the military and the private sector, known as the dynamic retention model (DRM). The DRM is a structural, stochastic, dynamic, discrete-choice model of individual behavior in which individuals make retention decisions under uncertainty and have unique (or *heterogeneous*) tastes. In the model, civil service employees make decisions throughout their careers about whether to stay in the civil service or leave and enter the private-sector labor force. The taste factor captures individuals' preferences for DoD civil service relative to the external market and includes persistent nonmonetary and monetary factors not otherwise included in the model. Individuals in the model are forward-looking, i.e., expectations about future events are incorporated into their decisionmaking, and decisions at a point in time are affected by the individual's employment history.

We extended the DRM to federal civil service employment and developed computer programs to estimate the model, using 24 years of data on defense civil service employees, and to simulate the effects of pay freezes and unpaid furloughs. The data are longitudinal and track the individual employment histories of the 1988 DoD entry cohort through 2012.

We demonstrate the DRM capability to study federal employee retention by estimating the model on subsets of General Schedule (GS) DoD employees, focusing on the most educated workers (those with a baccalaureate or higher degree) and on Science, Technology, Engineering and Mathematics (STEM) workers in the GS workforce.

Results

For each subgroup of GS employees for which we estimated our model, the fit of the model to the actual data is excellent, and all of the model parameter estimates are statistically significant. Interestingly, the mean of the factor representing individual heterogeneity—which we refer to as *tastes for public-sector employment in the DoD* relative to external private-sector opportunities—is positive for the subgroups we considered. That is, civilian employees, on average, have a positive taste for defense employment. The positive taste is unsurprising, in view of the fact that employees stay even though average federal pay for most of the subgroups we considered is less than average private-sector pay. They stay because they value the nonmonetary aspects of public employment (such as job security, stability, predictability, and being able to serve the public good).

Between 2011 and 2013, Congress froze federal civil service pay, resulting in no pay increase for three years. Federal civil service employees would have expected to receive a 1-percent per year pay raise, given the changes in the Employment Cost Index (ECI), the index used to set civil service pay raises, according the Federal Employees Pay Comparability Act of 1990. We simulated the effects of this pay freeze on civil service retention in both the steady state and the transition period to it, considering a 1-percent pay freeze in each of the three years. We also simulated the effect of an unpaid furlough in one year, equal to a 3-percent pay cut. In the steady state, all workers who were employed at the time of the pay freeze left service. Given that a civil service career can span 40 or more years, it can take more than 40 years for the steady state to occur, hence the importance of also simulating the retention effects during the years of transition to the steady state.

We first considered the case in which the cut in pay associated with the pay freeze is permanent, i.e., Congress never restores pay to the pre-freeze levels (in real terms). We found that the permanent 1-percent per year pay freeze over three years decreases the size of the retained GS workforce with at least a baccalaureate degree by 7.3 percent in the steady state and the size of the comparable GS STEM workforce by 8.5 percent.

In the transition years, the effects on retention unfold slowly. That is, the reduction in retention as a result of the permanent pay freeze does not occur all at once but cumulates over time. This occurs because the retention impact is greater in the mid-career years (i.e., for those not yet eligible for immediate retirement), and it takes time for those who were junior at the time of the pay freeze to reach this phase of their careers.

We also considered the case in which the pay cut associated with the pay freeze is temporary, i.e., Congress restores pay to its pre-freeze trajectory immediately following the pay freeze, although it does not provide back pay for the loss during the freeze. In this case, we found that the pay freeze has virtually no impact on retention in either the steady state or the transition period, because in the model, individuals are forward-looking and the lost back pay is small. Although the workers make their retention decisions under uncertainty, they anticipate the full restoration of pay immediately after the freeze, so they do not alter their retention behavior.

More realistically, federal employers may have some uncertainty about whether Congress will restore pay and when. If we assume that employees think that Congress will eventually restore pay sometime within the next decade, we can bracket the effect of this uncertainty by considering the case in which there is a ten-year delay between the pay freeze and the full restoration of pay and comparing the results to the case where pay is immediately restored. With this in mind, we considered the case of a ten-year delay before full restoration of pay, where employees know from the outset that pay will be restored.

We found no change in steady-state retention, even with the ten-year delay. The steady state occurs when all new entrants are under the restored pay regime, so their retention is unchanged relative to the base case. However, retention changes in the transition years. For example, we found that the retained GS workforce falls by 2.8 percent five years after the pay freeze and by 3.5 percent ten years after the pay freeze. These declines reverse after the trajectory of pay is restored, although it takes time for the workforce to return to the baseline levels. Thus, 20 years after the pay freeze, the GS workforce that is retained is still 1.4 percent below baseline, despite the restoration of pay after the ten-year period. The pattern of effects is similar for the GS STEM workforce. These results are different from those in the case in which pay is immediately restored, where the change in retention is virtually zero. Thus, uncertainty over when pay will be restored can potentially have a large effect on retention, even if employees are certain that pay will be restored within the next ten years.

We also considered the case of an unpaid furlough, equivalent to a one-time 3-percent pay cut. We found no discernible effect on retention for the subgroups we considered, in either the steady state or the transition period. It is likely that the retention effect is even smaller than we found (and we found no discernible effect), because we did not account for the value of leisure associated with the furlough. Although our analyses of furloughs and pay freezes do not account for changes in expectations workers may have about the possibility of future freezes and furloughs, such changes could have a negative, or even a positive, effect on retention.

Policy Implications and Directions for Future Research

Our analysis simulated the supply response of federal employees to compensation changes; however, it does not indicate the supply response relative to requirements. Put differently, our simulations were performed under the assumption that current pay levels were adequate but not excessive. If federal pay is in fact higher than necessary to sustain retention to meet workforce requirements, the decrease in pay associated with a pay freeze may not affect DoD's ability to retain the workforce required. Thus, while our simulations indicate that a pay freeze adversely affects the size of the civil service workforce that is retained, they do not necessarily imply that federal pay should therefore be increased or restored.

Assessments of the adequacy of federal pay levels must consider current and future workforce requirements as well as supply. If planners deem the supply to be inadequate, the DRM can provide empirically based simulations of the impact on retention of alternative compensation policies aimed at increasing supply. Similarly, if planners want to change the experience mix of the workforce, the DRM can help find temporary and permanent changes to the compensation structure that move the experience mix to the new target rapidly and cost-effectively. DRM policy simulations can show the year-to-year evolution of the experience mix of those retained and the annual cost of the policy under consideration.

There are a number of fruitful areas for future research. The DRM could be extended to simulate the retention effects of other policies of interest, including the effects of recent proposals by lawmakers to increase federal pension contributions, eliminate the Federal Employees Retirement System (FERS) annuity supplement, lower the FERS basic plan multiplier, and use the so-called chained Consumer Price Index to calculate cost-of-living adjustments for federal retirees. It could also be extended to assess the effects of incentive pays targeted on the retention of key subgroups of personnel and the effects of severance pay, early retirement, and other drawdown policies to induce voluntary separations.

The DRM capability could also be extended to other occupational areas within DoD, including the cyber workforce; to other pay systems, such as the STEM workforces in the various demonstration programs; to specific demographic groups, such as women and minorities; and to specific locations of interest, such as Hawaii. Furthermore, with appropriate data, the DRM capability we developed could be applied to civil service workforces in other agencies within the federal government, including the Department of Veterans Affairs, the Department of Homeland Security, and the various agencies comprising the intelligence community.

Finally, the DRM could be extended to create a "total force" model of DoD retention dynamics and the effects of compensation on those dynamics, where "total force" includes active and reserve military personnel and DoD civilians (but not contractors). RAND has already developed a unified DRM capability to provide logically consistent and empirically based estimates of the effects of compensation policy on active-component retention and reserve-component participation; this capability could be extended to include DoD civil service employment. There is already a flow of ex-service members to DoD civil service employment, but the role of compensation structure and incentives in influencing this flow has not been studied.

Given the importance of the federal workforce, it is critical that planners and policymakers have the capability to understand how changes in compensation and personnel policy affect that workforce. The DRM can provide such a capability, and the analysis presented in this report represents a step toward it.

Acknowledgments

We would like to thank the Defense Manpower Data Center for providing data for this study. We are grateful to Rachel Louie, formerly of RAND, and Beth Roth at RAND for outstanding research programming support. They both did a terrific job in constructing the data file we used. We appreciate the thorough reviews provided by our reviewers, Shanthi Nataraj at RAND and Justin Falk of the Congressional Budget Office. We would like to thank John Winkler, Director of the Forces and Resources Policy Center within RAND, for his support of this effort.

Abbreviations

ACOL	annualized cost of leaving
ACS	American Community Survey
BA	Bachelor of Arts
CBO	Congressional Budget Office
CSA	combined statistical area
DMDC	Defense Manpower Data Center
DoD	Department of Defense
DRM	dynamic retention model
ECI	Employment Cost Index
FEHB	Federal Employees Health Benefits
FERS	Federal Employees Retirement System
GS	General Schedule
MA	Master of Arts
MRA	minimum retirement age
MSA	metropolitan statistical area
NCS	National Compensation Survey
PhD	Doctor of Philosophy
SE	standard error
STEM	Science, Technology, Engineering and Mathematics
TSP	Thrift Savings Plan
VERA	Volunteer Early Retirement Authority
VSIP	Voluntary Separation Incentive Payment

CHAPTER ONE

Introduction

Congress froze federal civil service pay between 2011 and 2013, giving no pay increase for three years. Federal civil service employees would have expected to receive a 1-percent per year pay raise over this time period, given the changes in the Employment Cost Index (ECI), the index used to set civil service pay raises, according to the Federal Employees Pay Comparability Act of 1990.[1] In addition, sequestration in 2013 led to unpaid furloughs for some civil service workers, including those at the Department of Defense (DoD). These pay actions on top of a wave of baby-boomer retirements and various proposals by lawmakers to reduce federal compensation have raised concerns about the ability of federal agencies to sustain their workforces and attract and retain personnel in sufficient numbers, especially in critical skill areas.

Unfortunately, planners and policymakers in the federal government have little capability for assessing how changes in pay and unexpected unpaid furloughs will affect federal civil service retention and, specifically, the size and experience mix of the civilian workforce. What is required is a model of individual civil service employees' decisions to stay or leave the agency, estimated with individual-level data, that enables simulations of how changes in federal pay and personnel policy affect retention.

While there a number of studies comparing federal and private-sector pay levels,[2] research is lacking on how changes in pay and personnel policies affect retention behavior. As discussed in Chapter Two, past research is limited, and the studies that are available use a methodology that has distinct drawbacks, although it is easy to use. A methodology that avoids these drawbacks, known as stochastic dynamic programming, or the dynamic retention model (DRM), has been available for several decades, but its complexity and computational burden have often prevented analysts from implementing it. Fortunately, advances in technology have made the computational requirements far less demanding, and the improved methodology has now been used successfully in analyzing retention of military personnel as well as other workforces in the private sector.

The research summarized in this report was undertaken to apply the DRM approach to the federal civil service, and specifically to the DoD General Schedule (GS) workforce. We extended the stochastic dynamic model of retention to DoD civilians, estimated the model with longitudinal data spanning 24 years for an entry cohort of DoD civil service employees, and developed a simulation capability that permits us to simulate the effects of different compensation policies. We analyzed the retention behavior of key subgroups within the GS workforce, namely the Science, Technology, Engineering and Mathematics (STEM) workforce, as

---

[1] The President may limit the annual increase by executive order (Purcell, 2010).

[2] See, for example, Biggs and Richwine (2011), Edwards (2012), and Falk (2012).

well as the GS workforce that has at least a baccalaureate degree. The methods we used are general and can be applied to other civil service workforces, including those outside DoD.

While the research focuses on model development, we demonstrate the capability by considering two recent policies affecting federal civil service: the three-year pay freeze and the unpaid furloughs in 2013. Our analysis provides quantitative estimates of the effects of recent pay freezes and furloughs on civil service retention within a federal agency. We first considered the effect on retention of a three-year pay freeze of 1 percent per year, ultimately resulting in a permanent 3-percent drop in civil service pay. The three-year pay freeze of 1 percent per year is consistent with the actual pay freeze that occurred between 2011 and 2013 in the federal civil service.

Congress might ultimately restore civil service pay to its pre-freeze trajectory, that is, the levels of pay that would have resulted had the freeze not happened. Therefore, we also considered the retention effects of two cases in which Congress restores pay to pre-freeze levels going forward into the future but does not provide back pay for the three years of lost pay raises. In the first case, Congress is assumed to restore federal pay immediately after the end of the freeze; that is, pay levels are restored in the fourth year. In the second case, we assume that employees think that Congress will eventually restore pay some time within the next decade but are not sure exactly when. We bracketed the effect of this uncertainty by considering the case in which there is a ten-year delay between the pay freeze and the full restoration of pay and comparing the results to those of the case in which pay is immediately restored.

This bounding analysis is, of course, imperfect, because it does not directly model uncertainty over the year when pay will be restored. Although employees are certain that pay will be restored after ten years, uncertainty still enters their decisionmaking process via an annual environmental disturbance, or "shock," that affects the relative attractiveness of DoD civil service and other alternatives. Employees anticipate the shock but do not know the exact size of it until it is realized. Shocks will affect an individual's decision to stay or leave during the time between the start of a pay freeze and when pay is finally restored. It is also imperfect because we did not incorporate the possibility that employees revise their expectations about future pay actions and may believe that future pay freezes are more likely. We describe our policy analysis and results in more detail in Chapter Four.

Chapter Two describes the DRM as applied to civil service personnel. It provides a review of past quantitative studies of civil service retention, and it presents the theoretical model, along with the data we used and the specific subgroups of DoD civil service employees we considered. Chapter Three presents the model estimates and the fit of the model to the data. Chapter Four presents our policy analysis and shows the simulated effects of the pay freezes and unpaid furloughs. Policy implications and areas for future research are discussed in Chapter Five.

CHAPTER TWO

A Stochastic Dynamic Model of Retention for DoD Civil Service Employees

The DRM approach has been used in recent years in economics and management[1] and more extensively in defense manpower to analyze retention of active- and reserve-component military personnel.[2,3] However, it has not been used to analyze and assess compensation policies in the civil service. This chapter describes a DRM of civil service retention. It begins with a review of past studies of civil service retention and recent studies applying the DRM approach to military personnel. It then presents the theoretical DRM for civil service employees. The data we used to estimate the model and the estimation methodology are described. Model estimates and fit are presented in Chapter Three.

Past Studies of the Effects of Pay Changes on Federal Civil Service Retention

Several studies have quantitatively assessed the characteristics of federal civil service personnel who quit versus the characteristics of those who stay or provided descriptions of retention over time,[4] but only two have provided quantitative estimates of how changes in federal civil service compensation affect separation behavior. Both studies analyze separations from DoD civil service. The first, Black, Moffitt, and Warner (1990a), focuses on separations among DoD GS civilians in three occupational areas—administration, technical, and scientists and engineers—using data on individuals who entered DoD civilian employment between 1974 and 1977, tracking their individual careers for up to nine years or until 1983. The second, Asch, Haider, and Zissimopoulos (2005), focuses on retirements among DoD GS employees, using data on individuals who turned 50 years of age between 1980 and 1985 with at least 15 years of service and covered by the Civil Service Retirement System, tracking their individual careers until separation or until 1996.

These studies share a number of features. They use longitudinal data on individual civilians to estimate a model of the responsiveness of separation to financial incentives; allow for

---

[1] See, for example, Hotz and Miller (1993), Rust (1994), Keane and Wolpin (1997), Bajari, Benkard, and Levin (2007), Aguirregabiria and Mira (2010), van der Klaauw (2012), and Borkovsky, Doraszelski, and Kryukov (2012).

[2] See, for example, Asch, Hosek, and Mattock (2013), Asch et al. (2008), Mattock, Hosek, and Asch (2012), Mattock and Arkes (2007), and Gotz and McCall (1984).

[3] The first recorded application of this methodology was a study of U.S. Air Force officer retention in Gotz and McCall (1984), as acknowledged in Rust (1994).

[4] See, for example, Gates et al. (2013), Gates et al. (2008), Asch (2001), and Asch and Warner (1999).

heterogeneity among individuals, i.e., allow individuals to differ in their tastes for civil service relative to their external opportunities; and model the financial incentive to stay for another year relative to the incentive to leave as being equal to what is known in the literature as the annualized cost of leaving (ACOL).

In its simplest form, ACOL is the difference in the present value of the income stream to be obtained from leaving immediately and the income stream from staying *s* more years in the civil service, on an annualized basis. Thus, if *s* is 5, ACOL is the annualized difference between the present value of leaving immediately and receiving external private-sector pay and any possible future retirement benefits for which the individual qualifies, given departure at that time, and the present value of the stream of civil service pay over the next five years plus the present value of any future retirement benefits for which the individual might be qualified conditional on staying five more years. Similarly, if *s* is 10 or some other number, say 30, ACOL is the annualized difference when the time horizon is 10 or 30 years.

Under the assumption that the most relevant horizon is the one that minimizes the income loss from leaving immediately, the researcher chooses *s* that minimizes the annualized cost of leaving the civil service. Put differently, the ACOL is the annualized difference between the expected value of leaving immediately and the expected value of staying in the civil service for *s* years, where *s* is the horizon at which this annualized difference is at a minimum. This value of *s* is the optimal number of years an individual should stay in the civil service, i.e., the optimal quitting date. For example, in the context of retention of military personnel, where the ACOL modeling approach has been used extensively, ACOL is often minimized at *s* = 20, i.e., when military personnel are vested in their retirement system. Researchers, including the authors of the two studies of civil service retention, use the ACOL value in their models of the effects of changes in pay on retention.[5]

Black, Moffitt, and Warner (1990a) found a positive and statistically significant relationship between the ACOL variable and the quit behavior of administrative and technical personnel. That is, the higher the financial returns to leaving relative to those of staying, the higher the tendency to quit federal service. But contrary to their expectations, they found a negative—but not statistically significant—relationship for scientists and engineers (that is, the higher the financial returns to leaving relative to those of staying, the *lower* the tendency to quit federal service). Black, Moffitt, and Warner used their estimates for administrative and technical personnel to simulate the effects on quit behavior during the first nine years of federal service of a 10-percent permanent pay cut. They found that such a reduction would increase annual quit rates during the first nine years in these occupations by about 9 percent among technical workers and about 4 percent among administrative workers. Cumulative retention over the first nine years of service is estimated to fall by 3.4 percent for administrative employees and 7.8 percent for technical employees. Asch, Haider, and Zissimopoulos (2005, 2009) found a statistically significant effect between the ACOL variable and separation among

---

[5] The ACOL technique can be viewed as a way of approximating the value associated with making future decisions, given new information. This value is properly calculated as the expected value of the maximum over all the future decisions that might be available to an individual. The ACOL technique approximates this by computing instead the maximum of the expected values; while this approximation can, in practice, be computed rapidly, it has the disadvantage of implicitly ignoring any value associated with future flexibility in decisionmaking, as discussed later in the text. Thus, the value given by ACOL approximation is always less than or equal to the true value given by the expected value of the maximum. (More formally, this is a direct consequence of Jensen's inequality, which implies that $E[\max(x,y)] \geq \max(E[x],E[y])$.) The DRM uses the exact calculation.

retirement-eligible civil service employees that is consistent with Black, Moffitt, and Warner's earlier findings for administrative and technical personnel. They found that a $10,000 increase in the value of staying in the civil service relative to leaving is associated with a 4-percent lower probability of retiring from civil service.

While both studies suggest that civil service retention is responsive to financial incentives, the ACOL approach has important theoretical and practical drawbacks. Gotz's reply to Black, Moffitt, and Warner (1990a), published in the same issue of the journal, summarized both types of drawbacks (Gotz, 1990). From a theoretical standpoint, the ACOL approach is inconsistent with forward-looking rational behavior on the part of the civil service decision-maker, because it does not systematically include future random outcomes in the decision-making calculus. In the ACOL approach, random shocks do not enter the computation of the ACOL value, so the return to staying in the civil service does not depend on random shocks. In addition, the anticipated optimal career length does not depend on random shocks, because it is a direct function of the ACOL. The optimal career length is the length s that minimizes the value of the ACOL. Thus the ACOL model assumes that decisionmakers behave as if they know with certainty when they will leave. They are repeatedly surprised by random factors in each future period, even though random factors always occur.

From a practical standpoint, the ACOL approach can lead to implausible predictions about the retention effects of certain pay policies, thereby leading to flawed policy recommendations. Any policy that does not affect the minimum ACOL will not affect predicted retention in the ACOL model, even though common sense would suggest that such policies could do so. For example, in the ACOL model, a pay freeze could have no effect on retention if it does not affect the optimal ACOL or time horizon s, despite the fact that one would expect a pay freeze to have some impact on the retention of some individuals. Indeed, as we show in Chapter Four, our model provides evidence that a pay freeze could affect civil service retention.

Another example is the case of fixed contract lengths. Fixed contract lengths eliminate individuals' flexibility to respond to unexpected changes in their circumstances during the time of the contract. For example, if an individual signs a four-year contract, he or she no longer has the flexibility to respond to unexpected changes during the four-year contract period, without a significant cost for breaking the contract. One would expect that the lack of flexibility would reduce retention unless the individual is offered an additional incentive. Indeed, contracts are usually associated with a financial incentive to offset the negative effect of the lack of flexibility. The ACOL model assumes that members do not value such flexibility. Thus, imposing an employment contract, as often happens when employers offer an incentive such as a scholarship program, would have no effect on retention. Policymakers guided by the ACOL model would therefore tend to underestimate the financial incentive required to sustain or increase retention when they offer such contracts.

A final example is the transition period when a new compensation policy is introduced. During the transition period, currently serving employees are often grandfathered under the existing policy, and only new employees are covered by the new compensation policy. Currently serving members may be offered a choice to opt into the new policy, as was the case during the transition from the Civil Service Retirement System (CSRS) to the Federal Employees Retirement System (FERS). CSRS employees were given an open enrollment period when they could opt into FERS. The problem with the ACOL model, as explained by Gotz (1990), is that it cannot predict the proportion who would choose to stay under the old policy versus the proportion who would opt into the new policy. Comparable individuals at a given point in their careers would compute the ACOL under the old and new retirement systems, and all would choose the one with the higher annualized cost of leaving.

The DRM approach addresses these drawbacks. Black, Moffitt, and Warner (1990b) agreed that the DRM approach was in principle a preferable methodology in their reply to Gotz's reply (also published in the same journal issue), but they argued that, in practice, the computational burden made the DRM approach infeasible and that its use would have to await technological improvements in computational speed. Such improvements have occurred since 1990, and the DRM approach has been increasingly used in economics, management, and defense manpower, as reflected in published studies.

The study reported here is the first to apply the DRM approach to civil service retention. The remainder of this chapter presents the theoretical model and discusses data and estimation. While the DRM has a clear advantage over other methods, it has its own limitations, at least as we have implemented it so far. For example, we assume civilian pay and external salary opportunities are time-stationary, i.e., we do not allow their real value to vary over time. The model is estimated with only one entry cohort, the 1988 cohort, and the estimates could differ if additional cohorts were used. We assume a constant personal discount rate and do not allow personal discount rates to vary with age for an individual or to vary across individuals. The model excludes covariates such as gender, marriage, and location and ignores health status and health-care benefits. We do not model the accession decision or the individual's timing of entry into the civil service, and we do not model possible reentry to the civil service by those who separate. We also treat a transfer to another federal agency as a separation. In our description of the model and the discussion of the model estimates in the Chapter Three, we discuss these limitations in more detail, their relevance and importance to our analysis and results, and areas for future research that would address them.

A Dynamic Retention Model of DoD Civil Service Retention

The DRM is an econometric model of retention behavior. In it, employees make retention decisions each year over their career with a given employer. The model assumes that these employees are rational and forward-looking, taking into account their expected future earnings from the employer, as well as their own preference for employment with that employer, and uncertainty about future events that could cause them to value their current service more or less, relative to their external opportunities. Once the parameters of the underlying decision process, described below, are estimated, we can use these estimates to simulate the baseline retention profile of an entry cohort of civil service personnel, as well as the retention profile under alternative compensation policies. By appropriately scaling the results, we can make inferences about the effect of those policies on the size of the workforce that is retained and the required accessions needed to sustain the workforce should it decrease. While we do not explicitly model accessions, accessions are a by-product of the analysis, because they provide an estimate of how much accessions must change to sustain a given workforce size, given changes in retention resulting from a policy change.

We used a simple version of the DRM in this analysis, and while the simple version has been described in a number of past studies (see, for example, Mattock and Arkes [2007]), it has not been described in the context of the civil service, so we lay it out in some detail here for readers who are unfamiliar with those other studies. The discussion gives a broad overview, followed by the technical details of the model.

We modeled civil service retention from the start of employees' careers there. While we did not model the decision to enter the civil service and specifically when to enter it, employees in the model can enter the civil service for the first time at any age. In the data we used here, civil service employees enter between the ages of 22 and 52.

Each year, the individual compares the value of staying in civil service with that of leaving and bases his or her decision on which alternative has the maximum value. In the basic DRM, we assume that once individuals leave DoD civil service, they do not reenter at a later date.[6] Individuals who stay can revisit the choice between DoD civil service employment and external opportunities in each future period until retirement from the labor force, which is presumed to be age 65. All of these decisions will depend on the employee's unique circumstances at a given point in time. Those circumstances include preference for DoD civil service relative to external opportunities and random events that may affect relative preferences.

In the model, the value of staying depends on the annual civil service earnings in each time period. Annual earnings vary with age, and those who begin their civil service career at older ages also begin at higher pay than their younger counterparts.

The value of staying also depends on the individual's preference for DoD civil service relative to the external market (his or her "taste" for civil service), civil service compensation, and a period- and individual-specific environmental disturbance term (or shock) that can either positively or negatively affect the value placed on civil service in that period. For example, having an ailing family member who requires assistance with home care could be such a shock that may decrease the value placed on civil service employment. The taste for civil service is assumed to be constant over time for a given individual and can be thought of as the net effect of idiosyncratic, persistent differences related to the individual's perceived value of working in the civil service relative to the external market. The net effect includes all nonmonetary and monetary factors the individual perceives as relevant to the civil service over and above monetary factors included in the model. These factors might include patriotism and desire for public service, positive and negative things the individual perceives about the civil service (e.g., hours of work, differences in paid leave), and persistent differences in civil-service and private-sector earnings apart from the differences accounted for in the model. As mentioned, we use a single curve to represent GS salary and external salary by age. But an individual might believe his or her GS and external salaries are persistently higher or lower than those curves. The net effect of these perceived differences would enter into taste. Another way of describing taste, then, is as a person-specific fixed effect.

Individuals are heterogeneous with respect to their tastes for civil service, i.e., their tastes differ. As we discuss below, we as analysts did not directly observe these tastes, but we assume they are distributed according to a known type of probability distribution but with unknown parameters. A goal of the estimation process is to estimate these parameters.

---

[6] This, in fact, is not true. Civil service employees can flow in and out of DoD civil service. Furthermore, the DRM can accommodate such flows, as is done in Asch, Hosek, and Mattock (2013) and earlier studies that permit flows of military personnel in and out of the reserve components. However, the data we use are for the new-entrant cohort who entered DoD civil service in 1988, and only 9 percent of them ever return between 1988 and the end date of our data, 2012. We therefore exclude this possibility (and these observations) for the purposes of this study.

The value of staying in the civil service also includes the value of the option to leave at a later date. That is, the individual knows that he or she can revisit and reoptimize the decision to stay or leave in the next period. Of course, the future is uncertain, so the value of being able to choose to stay or leave in the future is expressed as the discounted present value of an expected value. Individuals may reoptimize, and might change their decision in the future because new information, e.g., a new shock, makes it reasonable to do so or because the discounted expected value of future benefits of leaving becomes greater relative to the benefits of staying. Furthermore, choices made today can affect the value of choices in the future. An individual who chooses to stay in the civil service today adds a year of service, moving closer to retirement eligibility and increasing retirement benefits, thereby influencing the value of choosing the civil service in the future. Similarly, past choices can affect the value of current and future choices.

The value of leaving includes the value of the external alternative, which includes pay in the external market, any civil service retirement benefits the individual is entitled to receive, and an individual- and period-specific shock term that can either positively or negatively affect preference for the external alternative. Pay in the external market varies with age, with those entering the civil service at older ages having higher external pay opportunities. Entry age can also affect civil service retirement benefits.

An individual who leaves DoD may either transfer to another federal agency or leave federal service altogether. Thus, the value of leaving could include federal pay and expected federal retirement benefits in the next federal job. In the data we used to estimate the model, about 10 percent of DoD separations within the first nine years of service are transfers to other federal agencies. With data that track the careers of DoD civilians both within the DoD civil service and in other federal agencies, we can easily extend the DRM to incorporate transfers. However, with the available data, we do not know how long transferees stay in federal employment if they are not in DoD. Consequently, in the current analysis, we assume all DoD separations are separations from federal employment, although we recognize that this will lead to some measurement error of the value of external opportunities for those who transfer.

More technically, we can write the value of staying and the value of leaving in each time period, as was done in past studies. The value of staying in DoD civil service is

$$V_t^S = \gamma^c + w_t^c + \beta E_t \left[Max\left(V_{t+1}^S, V_{t+1}^L\right)\right] + \varepsilon_t^c \qquad (2.1)$$

where

V_t^S is the value of staying in the DoD civil service at time t

γ^c is individual taste for DoD civil service relative to the external market

w_t^c is civil service annual earnings at time t

β is the civil service employee's personal discount factor

V_{t+1}^S is the value of staying in DoD civil service at time $t + 1$

V_{t+1}^L is the value of leaving DoD civil service at time $t + 1$, defined in Equation 2.2

$E_t \left[Max\left(V_{t+1}^S, V_{t+1}^L\right)\right]$ is the expected value of having the option to choose to stay or leave in the next period

ε_t^c is the random shock to DoD civil service employment at time t.

Similarly, the value of leaving DoD civil service at time t is

$$V_t^L = w_t^e + \sum_{s=t+1}^{T} \beta^{s-t} w_s^e + R_t^c + \varepsilon_t^e \qquad (2.2)$$

where

V_t^L is the value of leaving DoD civil service at time t

w_t^e is annual earnings in the external market at time t plus retirement benefits that will accrue to the civil service employee in the external market from t until T

$\sum_{s=t+1}^{T} \beta^{s-t} w_s^e$ is the present value of future external market earnings

R_t^c is the retirement benefit accrued as a result of civil service employment for an individual leaving at time t (exclusive of any benefit accrued from work in the external market)

ε_t^e is the random shock to external employment at time t.

We assume that to claim civil service retirement benefits, the individual must have left the civil service. That is, we do not include the possibility that a civil service employee may become a reemployed annuitant and claim retirement benefits while working in the civil service. We excluded this possibility to simplify our empirical analysis, although the DRM can be extended to allow the choice to become a reemployed annuitant. As we describe below in our discussion of data, our empirical analysis focuses on DoD civilians covered by FERS. The FERS benefit also includes Social Security, and we included the formulas for the Social Security benefit in 2011 in our empirical model. We assume civil service employees who separate take employment that is also covered by Social Security. Thus, when we compute the expected retirement benefit under FERS as part of R_t, we net out the expected Social Security retirement benefits the employee might qualify for as a result of his or her external market employment. Private-sector retirement benefits are absorbed into the taste term.

The individual is assumed to decide to stay in DoD civil service at time t if the value of staying is greater than the value of leaving, or

Stay at time t if V_t^S = max (V_t^S, V_t^L)

Thus, the probability of staying at time t is

$$Pr_t(Stay) = Pr\left(V_t^S > V_t^L\right) = Pr\left(E\left(V_t^S\right) - E\left(V_t^L\right) > \varepsilon_t^e - \varepsilon_t^c\right) \qquad (2.3)$$

Referring back to Equations 2.1 and 2.2, we see that the current wage enters the value function linearly and has a coefficient of one. However, the decision to stay depends not only on the current wage but also on the value of the entire value function, which also incorporates taste, current shock, and the expected value of the maximum in the next period. Although the model's structure may seem simple because the current wage enters additively, it is in fact complex, and the stay/leave decision depends on a full assessment of current and future opportunities. As shown below, the model fits civil service retention data well.

More-complex model specifications have been used in other work. For instance, dynamic programming has been applied to analyze retirement decisions and full- versus part-time work choices (van der Klaauw and Wolpin, 2008). Such models use a period-specific utility function, and the objective is to maximize intertemporal utility subject to initial assets, saving behavior, and the retirement system, e.g., Social Security. Such specifications are potentially useful for analyzing civil service retention or military retention, but available data limit what can be done. Data on spouse earnings, full- versus part-time work, savings, wealth, and the timing of retirement are absent, for example. Stated differently, our value-function specification can be thought of as a particular form of utility function in which current utility depends additively on the current wage, taste, and shock, plus the discounted expected value of following the best path in the next period.

We do not observe individuals' tastes for the civil service or random shock terms. Instead, we assume they are each distributed according to known types of probability distributions with unknown parameters that we estimate using available data. Specifically, we assume individuals' tastes for civil service are normally distributed and the random shocks have an extreme-value type 1 distribution. Given these distributional assumptions, we can derive choice probabilities for each alternative at each decision year and the cumulative choice probabilities or survival probabilities for an entering cohort at each decision year and then write an appropriate likelihood equation to estimate the parameters of the model. These parameters include the parameters of the probability distribution for the shock terms, the parameters for the population distribution of taste for civil service, and the discount factor.

We next derive the choice probabilities, the cumulative probabilities, and the likelihood equation. The extreme-value distribution, $EV[a,b]$, has the form $\exp(-\exp((a-x)/b))$ with mean $a+b\gamma$ and variance $\pi^2 b^2/6$, where γ is Euler's Gamma (approximately 0.577), a is the location parameter, and b is the scale parameter. We assume the shock terms have a zero mean and scale λ, implying that they have the extreme-value distribution $EV[-\gamma\lambda, \lambda]$, i.e., $a = -\gamma\lambda$ and $b = \lambda$. Since both ε_t^e and ε_t^c have an extreme-value distribution, the difference in Equation 2.3 is known to have a logistic distribution. With this information, the expected value of the maximum of V_{t+1}^S and V_{t+1}^L can be written as

$$
E_t\left[Max\left(V_{t+1}^S, V_{t+1}^L\right)\right] = \iint Max\left(V_{t+1}^S, V_{t+1}^L\right) d\varepsilon_t^c d\varepsilon_t^e = \lambda ln\left[e^{\frac{V_{t+1}^S}{\lambda}} + e^{\frac{V_{t+1}^L}{\lambda}}\right]
$$
(2.4)

Consequently, we can write the expected value of V_t^S as

$$
E\left[V_t^S\right] = \gamma^c + w_t^c + \beta\lambda ln\left[e^{\frac{V_{t+1}^S}{\lambda}} + e^{\frac{V_{t+1}^L}{\lambda}}\right]
$$
(2.5)

Thus, we have an explicit expression for the value function, given (unobserved to the analyst) taste for civil service, γ^c. (Later in this subsection, we describe how we handle unobserved tastes by integrating out this source of heterogeneity.) Given Equation 2.4, we can write the probability that a civil service employee chooses to stay at time t as

$$Pr_t\left(Stay\right)=\frac{e^{\frac{V_t^S}{\lambda}}}{e^{\frac{V_t^S}{\lambda}}+e^{\frac{V_t^L}{\lambda}}} \qquad (2.6)$$

The probability of leaving at time t is simply $1 - Pr_t(Stay)$.

Given these probabilities, we can write the cumulative probability that a civil service employee entering at time 1 will stay through t as

$$cumulativePr\left(Stay\right)_t=\prod_{s=1}^{t} Pr_s\left(Stay\right) \qquad (2.7)$$

The cumulative probability that a civil service employee stays for $t-1$ years and leaves at t is

$$cumulativePr\left(Leave\right)_t=\prod_{s=1}^{t-1} Pr_s\left(Stay\right)\left(1-Pr_t\left(Stay\right)\right) \qquad (2.8)$$

These probabilities are conditioned on the unobserved taste parameter, γ^c, since the value of staying, V_t^S, depends on γ^c. As mentioned, we assume the taste parameter has a normal distribution $g\left(\gamma^c\right)$ with mean μ and standard deviation σ. We use this information to formulate the expected cumulative probability of a given career path, or the likelihood of that path. Thus, for a civil service employee in our data who stays through $t-1$ and leaves at t, the likelihood of that career path is

$$L_i\left(\mu,\sigma,\lambda,\beta\right)=\int_{-\infty}^{\infty}\prod_{s=1}^{t-1}Pr_s\left(Stay\right)\left(1-Pr_t\left(Stay\right)g\left(\gamma^c\right)d\gamma^c\right) \qquad (2.9)$$

The subscript i in L_i denotes the ith civil servant in our data. Similarly, if the individual stays through t and is then censored, the likelihood is

$$L_i\left(\mu,\sigma,\lambda,\beta\right)=\int_{-\infty}^{\infty}\prod_{s=1}^{t}Pr_s\left(Stay\right)g\left(\gamma^c\right)d\gamma^c \qquad (2.10)$$

Thus, the likelihood for the entire data sample, N, is given by

$$L_i\left(\mu,\sigma,\lambda,\beta\right)=\prod_{i=1}^{N}L_i\left(\mu,\sigma,\lambda,\beta\right) \qquad (2.11)$$

Because we allow entry age to vary, the likelihood in Equation 2.11 also depends on entry age, although we do not explicitly include it as one of the arguments. The aggregate likelihood function is a weighted average of the likelihood-function values at each entry age, where the weights are the fraction of the entry cohort at each entry age.

Estimation

The parameters we estimated are the mean and standard deviation of the taste distribution, the location parameter of the shock distribution, and the discount factor. As we describe below, we estimated two additional parameters, the probability of attrition in the first year of DoD civil service employment and the probability of being censored in the ninth year (1997 in our data, when a reorganization resulted in some DoD employees being moved to an organization outside of DoD). Also, we emphasize that the model was estimated from actual data and is not calibrated. Calibration would select parameter values from a sequence of guesses that depend on model fit under prior guesses, whereas estimation finds the parameters that simultaneously maximize the model's fit to the data and provides standard errors of the estimates by which to judge their statistical significance.

The model's parameters were estimated with maximum likelihood, where the likelihood function is given by Equation 2.11. Optimization is done using the BFGS (Broyden-Fletcher-Goldfarb-Shanno) algorithm, a standard hill-climbing method. Standard errors of the estimates were computed using numerical differentiation of the likelihood function and taking the square root of the absolute value of the diagonal of the inverse of the Hessian matrix.

To compute the likelihood function in Equation 2.11, it is necessary to evaluate the integral in L_i, i.e., to integrate out the unobserved heterogeneity in taste for civil service employment. We did this by drawing 35 sample points that uniformly cover 99 percent of the normal cumulative probability distribution function defined using trial values of the taste distribution parameters. For each sample point, the dynamic program was solved for each individual, and the likelihood value for that individual was computed. We integrated over the distribution of tastes by taking the average of the likelihoods over the 35 sample points.

The process of estimation tries different values of the trial parameters until the career likelihoods are maximized for the sample of civil service employees used. While this is the standard approach in maximum-likelihood estimation, the computation burden associated with the DRM occurs because for each trial set of parameters, the dynamic programming problem has to be re-solved for all periods for all 35 draws of tastes for each individual. Then, given the new solution, the choice probabilities are updated, and the likelihood function is reevaluated to determine whether the fit has improved and in what direction the next trial parameters should be changed to improve it in the next iteration. Re-solving the dynamic program requires extensive computation for each individual in the data. Fortunately, because of tremendous strides in computational speed in recent years, estimation can be done using a laptop computer.

To judge goodness of fit, we used the parameter estimates to simulate retention rates by year of service of DoD civil service personnel and compared those rates to the actual data. We show goodness-of-fit diagrams in Chapter Three when we present the model parameter estimates. We next discuss our data and our simulation methodology.

Data

The data we used build on a longitudinal data file created for the analysis in Asch (2001) that drew upon Defense Manpower Data Center (DMDC) data on DoD civil service personnel. The DMDC data for that study included every GS employee in DoD civil service as of the beginning of each fiscal year and the transaction data indicating changes in each individual's personnel record, including whether he or she separated. The longitudinal data file constructed from these DMDC data tracked the individual careers of those who entered DoD civil service in 1988 through 1996.

In this study, we extended that longitudinal file to include more-recent information for each individual in the 1988 entry cohort through 2012. That is, we merged more-recent DMDC data on DoD civil service personnel with the 1988 cohort file so that we could track individual DoD civil service careers from 1988 through 2012 or separation, whichever came first. Thus, we were able to track the careers of DoD civil service personnel over a 24-year period. The data include 22,727 observations on GS workers who began their career in 1988.

We excluded some observations from the analysis: We excluded temporary workers and those who worked less than full time, were considered inactive or seasonal, or were military technicians, reducing the data to 19,713 observations. We also excluded individuals in the 1988 entry cohort who separated and later returned. (Only about 9 percent of those in that cohort returned.[7]) Excluding those who left and returned by 2012 (i.e., over the 24-year period) reduced the data to 17,899 observations.

All of the individuals in the data we used to estimate the DRM are covered by FERS, which was introduced in 1987. It consists of three parts: a defined benefit plan that bases retirement benefits on the employee's earnings and years of service, Social Security coverage, and a defined contribution plan called the Thrift Savings Plan (TSP). All new entrants with less than five years of service are automatically placed under FERS. Because the individuals in our analysis had no prior service in 1988, they fall into this category.

We estimated our model for four specific subgroups defined by education and, in one case, by occupation: (1) all GS workers with a Bachelor of Arts (BA) degree, a Master of Arts degree (MA), or a doctorate (PhD); (2) all GS workers with an MA or PhD; (3) all GS workers with a BA; and (4) GS STEM workers with a BA, MA, or PhD. We also estimated a different specification for GS STEM workers, so we estimated five models in total. More specifically, for the education subgroups, we estimated separate models for those with a BA and those with an MA or PhD (but not a professional degree). We also estimated the model for the GS STEM workforce, defined as GS employees with at least a BA and in occupation series defined as science, technology, engineering, or mathematics. We identified these subgroups on the basis of their education and occupation at entry. It is important to note that in this analysis, we considered only STEM employees who are part of the GS workforce. STEM employees in DoD are in a variety of pay plans other than the GS plan, including various demonstration projects. Thus, our results are relevant for only a subset of the broader DoD STEM workforce.

We recognize that individuals can gain additional education while they are in the civil service and can switch occupations. However, as discussed in Asch (2001), the civil service data do not appear to always update the education information as individuals obtain more education, so we opted to use the entry education level. While the DRM could be expanded to consider occupational choice, we do not pursue that avenue here, so we focus only on entry occupation. Although the DRM could be used to consider other subgroups, we focus only on the more-educated GS employees.

Civil Service Pay and Retirement Benefits and Private-Sector Wages

Important inputs to the DRM are civil service pay, w_t^c, civil service retirement benefits net of external retirement benefits, R_t, and external pay, w_t^e, We chose to estimate separate models

---

[7]   The percentage was even lower for the STEM entry cohort we analyzed—about 7 percent.

for different subgroups in part because civil service pay differs by subgroup (as shown in Asch, 2001) and private-sector pay differs by subgroup as well.

We estimated the DoD civil service pay profile over each subgroup's career, as well as the external private-sector pay profile for each subgroup, using the Bureau of Census' American Community Survey (ACS) for 2009 to 2011. All pay data are in 2011 dollars. We computed mean annual earnings for full-time, full-year workers ages 25 to 65 for each education subgroup and for the STEM subgroup. We define full-year as having 40 or more weeks of work annually and full-time as working at least 35 hours per week. DoD civil service earnings are approximated with annual earnings of federal workers, so the pay line in our analysis is not DoD-specific. We used the mean annual earnings of full-time, full-year private-sector workers for each subgroup for the external pay line.

A January 2012 Congressional Budget Office (CBO) report compared federal pay with private-sector pay and found that differences between them, accounting for differences in observed characteristics (e.g., occupation and age), depended on education attainment and on whether the comparison was of pay alone or of total compensation (pay plus benefits). Overall, CBO found that average federal compensation was higher than private-sector pay, except for individuals with a professional degree or doctorate. In addition, it found that average federal pay was about the same as private-sector pay or lower for individuals with a BA degree or higher. Biggs and Richwine (2011) found that federal pay exceeded private-sector pay, controlling for differences in observed individual characteristics, with a smaller premium for the most highly educated federal employees.

The pay profiles we developed for our analysis are not a result of a regression analysis that controls for an array of individual characteristics. Indeed, our study is not intended to make such pay comparisons. However, our pay profiles control for age and education (and for STEM workers, for occupation) and focus on full-year, full-time employees. The pay profiles we used are broadly consistent with the findings of the CBO study. That is, for STEM workers and workers with either an MA or a PhD, federal pay falls short of private-sector pay in our data. For those with a BA, federal pay is less than private-sector pay for younger employees but higher for older employees.

There are several relevant issues related to the pay profiles: locality adjustments to GS pay, the appropriateness of 2011 pay profiles for the 1988 entering cohort, the probable decrease in the value of health-insurance benefits for workers who retire from external positions, and the reduction in the DoD civilian workforce that occurred in the 1990s.

In estimating the model, we used federal civil service pay profiles by age for different education groups, estimated from the ACS for 2009–2011, corresponding to 2008–2010. We have a single pay profile for each education group, but since 1994, GS pay has differed by locality. The annual adjustment in GS pay has been based on a "base GS" adjustment and a locality-specific adjustment. In the Appendix, we show that although locality pay has created different pay levels geographically, in most (85 percent) of the localities, the GS salary is within plus or minus 5 percent of the nationwide average GS salary, and that this pattern has remained the same since the introduction of locality adjustments in 1994. Thus, the geographic differences are persistent but relatively tightly bunched, and these differences contribute to the variance of taste in our model. Beyond these persistent differences, there is some year-to-year variation across localities in the annual adjustment which can contribute to the shock variance, but the contribution will probably be minor because the year-to-year variation is small.

We estimated our model on the 1988 entering cohort of DoD civilian employees with no prior federal civil service experience, but we used wage profiles from 2008–2010. To address the question of whether the 2008-2010 wage profiles are appropriate for a cohort that entered nearly a quarter of a century earlier, we referenced data on wages for federal civil servants and for full-time wage and salary workers 25 years of age and older. From 1990 to 2012, inflation-adjusted federal civil service scheduled pay, including locality adjustments, grew by 11 percent (see Table A.1 in the Appendix). By comparison, median usual weekly earnings of full-time wage and salary workers 25 years and older fell by 15 percent for those with less than a high school diploma and by 6 percent for those who were high school graduates with no college; they fell by 7 percent for those with some college or an associates degree but rose by 9 percent for those with a BA or higher.[8] For workers with an MA or a PhD, the CBO study suggests an even higher percentage increase. Given our focus on civil service employees who have four or more years of college, the wage data indicate nearly the same growth in (inflation adjusted) civil service pay as for wage and salary workers, an 11-percent increase versus a 9-percent increase.

Assuming that our cohort of 1988 DoD civil service entrants was aware of these wage trends, our use of 2008–2010 pay profiles places external wages at a comparable level relative to what they would have been between 1988 and 2012. This is more likely to be the case for workers with four years of college than for workers with an MA or a PhD, for whom private-sector wage growth was probably higher than civil service wage growth. Overall, the wage trends suggest that the 2008–2010 wage profiles are a reasonable proxy for wage profiles as envisioned by the 1988 entrants, although they are, of course, not perfect.

Because the 2008–2010 wage profiles for workers with four years of college are about the same, compared to civil service wages, as they would have been in 1988 and during the years from 1988 to 2012, the estimate of average taste for civil service probably has little bias. However, it might be biased upward for workers with more than four years of college.[9] A more complete treatment of this issue would involve extending our model to include wage expectations, but that is beyond the scope of this work. The wage trends also suggest that the use of 2008–2010 wage data would probably not have been appropriate for workers with less than a college education; inflation-adjusted federal pay rose by 11 percent between 1990 and 2012, while wage and salary workers saw their pay fall by 6 to 15 percent.

The trend in private-sector health benefits is also relevant to consider. Health-insurance costs have increased, and employers have tended to decrease the scope of the health benefit or eliminate it entirely. This affects external compensation for current employment, as well as health insurance provided to retirees by external employers. In contrast, "federal retirees and their surviving spouses retain their eligibility for FEHB [Federal Employees Health Benefits] health coverage at the same cost as current employees."[10] The Office of Personnel Management provides online access to FEHB plans by state, with information on health-insurance premiums and the portions paid by the government and the insured. We compared these data with the Kaiser Family Foundation 2013 survey of employer-sponsored health programs and concluded that the plans offered through FEHB are similar in cost to those offered in employer-

---

[8]   Derived by the authors from Table 19 in U.S. Bureau of Labor Statistics (2013).

[9]   If the expected external wage is actually lower than that measured by our pay line, retention in civil service can be supported by a lower average taste for civil service.

[10]   U.S. Office of Personnel Management, undated a.

sponsored health programs (Kaiser Family Foundation, 2013). Thus, the FEHB plans do not appear to have a cost advantage; their premiums and the portion paid by the insured are similar to those of employer-sponsored plans. We did not, however, find information on health benefits among college-educated retirees, i.e., workers comparable in education to the DoD civil servants in our data. Still, it seems likely that the provision of health insurance as part of an employer-sponsored benefit package for current employees and retirees has decreased in prevalence and benefit scope. If so, health insurance provided to federal civil service retirees has become relatively more valuable. The omission of a downward trend in the value of external health-insurance benefits will tend to bias downward the estimate of mean taste. The external option will be increasingly less valuable than it appears to be, so a lower average taste for DoD civil service will be sufficient to support retention. We do not know the size of this bias. If the downward trend in external health benefits has had relatively less impact on college-educated workers than on workers in general, the bias might be small.

Another relevant pay issue is the availability of Voluntary Separation Incentive Payments (VSIPs) and Volunteer Early Retirement Authority (VERA) to civil service personnel. DoD decreased its civilian workforce in the 1990s as part of the overall drawdown in defense manpower. The active components decreased from about 2.0 million members in 1990 to 1.4 million at the end of the 1990s, and the DoD civilian workforce decreased from about 1.0 million in 1990 to 800,000 in 1996 and 675,000 in 2000.[11] According to Peters (1996), "By the end of 1995, 79,000 Defense civilians had taken advantage of separation incentives." The incentives were VSIP, a one-time payment of up to $25,000 for eligible civilians, and VERA, which allowed civilians to retire early at any age if they had 25 years of service or were at least age 50 years old with 20 years of service.[12] VERA was often offered together with VSIP. Thus, the incentives encouraged early retirement. At the same time, new hiring of civilians largely ceased. By the 1990s, the 1988 cohort used in estimating our model had become permanent workers and probably faced little risk of being induced to leave. Therefore, we think the drawdown policies at that time had little effect on retention behavior on the cohort we consider and so had little effect on our estimation.

For retirement benefits, the FERS benefit formula was programmed into our model. An element of that formula is the minimum retirement age (MRA), which depends on birth year. In our analysis, we assumed MRA was 56 (the employees were born between 1953 and 1964). While the TSP includes a formula for DoD matching contributions that is based on employee contributions, we assumed for simplicity that DoD contributes 5 percent of each employee's annual earnings, the maximum DoD matching rate. We also assumed an annual real return on TSP investments of 5 percent. We did not include employee contributions as part of the retirement benefit, because these contributions are not part of federal compensation but are individual savings.

The FERS benefit also includes Social Security, and we included the formulas for the Social Security benefit in 2011 in our programming. However, we assumed that civil service employees who separate take employment that is also covered by Social Security. Thus, when we computed the difference in retirement benefits associated with Social Security, we included only the increment in Social Security benefits—specifically, the FERS Special Retirement Sup-

---

[11] DMDC, undated.

[12] "Employees under 55 receive a 2 percent deduction in their retirement annuities for every year they are under 55" (Peters, 1996).

plement. FERS includes a supplement for those who have completed a full career and leave before age 62. For example, FERS allows individuals who have served 30 years to retire at their minimum retirement age (assumed to be 56 in our model). The supplement provides a reduced Social Security benefit between the age at which they retire (e.g., 56) and age 62, the earliest age for Social Security retirement benefits. The Social Security Administration does not pay this benefit; the Office of Personnel Management pays it.

Accounting for Censoring and Early Attrition

Exploratory data analysis and early estimation results showed two empirical regularities for the subgroups we considered. First, we found a sizable drop in retention in the first year of service. For example, 15 percent of the entry cohort of personnel with a BA left before completing their first year of service. Second, we found a sizable drop in retention in the ninth year of service (corresponding to 1997) for the more highly educated groups, e.g., those with a BA or a higher degree. Further investigation showed that the drop corresponded to the disappearance of personnel in some occupations, especially cartography, from DoD civil service in 1997. This occurred because of a reorganization that caused some DoD functions to be exported to the newly created National Imagery and Mapping Agency (NIMA), a major employer of cartographers; various defense functions, including the Defense Mapping Agency, were combined.

We initially estimated models in which we excluded cartographers and ignored the issue of early attrition. (The results of this specification for the GS STEM workforce are shown in the first column of Table 3.1 in Chapter Three and are discussed further there.) We then estimated models in which we included these observations and developed a methodology for incorporating the issue of censoring in our data.

We handled the issue of early attrition and censoring at year 9 (in 1997) by incorporating two additional parameters in our model, δ and θ. Both parameters shift the probability of staying given in Equation 2.6 and therefore the cumulative probability of staying given in Equation 2.7.

To capture the shift due to dropouts in the first year, we write the probability of staying at $t = 1$ as

$$Pr_1(Stay)=(1-\delta)\frac{e^{\frac{V_1^S}{\lambda}}}{\left[e^{\frac{V_1^S}{\lambda}}+e^{\frac{V_1^L}{\lambda}}\right]}$$

To capture the shift in year 9 associated with reorganization, we write the probability of staying in $t = 9$ as

$$Pr_9(Stay)=(1-\theta)\frac{e^{\frac{V_9^S}{\lambda}}}{\left[e^{\frac{V_9^S}{\lambda}}+e^{\frac{V_9^L}{\lambda}}\right]}$$

For all other values of t, the probability of staying is given in Equation 2.6.

We estimated two models that include both δ and θ. The estimation results are shown in Chapter Three.

Simulation

We used the estimated parameters to simulate the cumulative retention probabilities at each year of service under the current compensation system (the baseline) and alternative policies. The simulation capability is quite general and can be used to assess a wide range of compensation changes, including targeted pays, such as retention bonuses or separation pay; changes to the retirement system; and changes to the level and/or structure of current pay. We demonstrated the capability by simulating the effects on retention probabilities of a temporary pay freeze of 1 percent annually over a three-year period. We also considered the effect of a furlough that results in a pay cut of 3 percent in one year. We simulated separate cumulative probability profiles for each entry age and then computed an average retention profile weighted by entry-age frequency observed in the data.

The simulations are based on the two models in which we included the additional parameters to account for early attrition and for censoring of some observations at year 9. Because the censoring was a result of administrative reorganization during the time period of our data and not a general phenomenon, we set the value of the censoring variable to zero in our simulations. That is, the simulations are based on estimates that account for censoring, but no observations are censored in them.

We evaluated the effects of a temporary pay freeze by simulating the year-by-year effects on the cumulative retention profile. That is, we simulated retention in each future calendar year over a 40-year period. This allows us to understand the short- and long-run effects of a policy. In the case of a pay freeze, the policy may have effects beyond the years when pay is frozen if pay is not later restored.

To simulate the effect of a policy that is not permanent, such as a three-year pay freeze, we needed to add another time clock in our model. The model already included a clock that counts years of DoD civil service, and it implicitly included a clock that counts age or total time since entering DoD civil service. To consider the transition period after a policy is introduced and to consider policies that are targeted to specific calendar years, we needed a clock starting at a civil service member's years of service when the policy change occurs, i.e., the member's cohort. For example, a three-year pay freeze is targeted to cohorts in service during the course of the pay freeze, but not earlier or later cohorts. Once we identified the cohort (year of service at the time of policy change), we used the DRM parameter estimates to simulate the DoD civil service retention behavior for each cohort over its entire career (up to age 65) and used the cohort-specific retention behavior to simulate the aggregate retention behavior of all cohorts for each time period since the policy change occurred. That is, we can show retention rates after a policy change by calendar year, just as a force planner might wish to see it. The methodology for introducing cohorts into the DRM and considering the transition phase is described in Asch, Hosek, and Mattock (2013), where the method is applied to the retention behavior of active-duty personnel. We extended this methodology to DoD civil service in this study.

We simulated the cumulative retention probabilities at each year of service in the baseline, without the policy change and then as a result of the policy in the steady state and in the transition. We rescaled the profiles to the fiscal year 2011 end strength. At the end of fiscal year 2011, the DoD GS STEM workforce with a BA or higher degree was 57,440, while the end strength for the DoD GS workforce with a BA or more was 225,888. This rescaling allows us to present the simulation results in terms of the size of the workforce. We chose 2011 because our pay profiles are in 2011 dollars.

CHAPTER THREE

Estimation Results

We estimated the DRM for subgroups of DoD GS civil service employees, namely, GS STEM workers with at least a BA degree and DoD GS employees with at least a BA. The models include the two additional parameters that capture early attrition and censoring in year 9. We also show results of an estimated model for GS STEM workers with at least a BA in which we do not include these additional parameters. Table 3.1 shows the parameter estimates and standard errors (SEs) for each model. Figures 3.1 through 3.4 address model fit. We first discuss the model fits and then specific parameter estimates.

Table 3.1
Parameter Estimates and Standard Errors for GS DoD Civil Service Subgroups

Parameter Estimate/ Standard Error	GS STEM Workers with BA, MA, or PhD (Model 1)[a]	GS STEM Workers with BA, MA, or PhD (Model 2)[a]	All GS Workers with BA only (Model 3)	All GS Workers with MA or PhD (Model 4)	All GS Workers with BA, MA, or PhD (Model 5)
Mean of the taste distribution, μ	9.14	11.20	10.53	20.66	13.97
SE	2.67	2.41	1.67	2.14	1.95
Variance of the taste distribution, α	25.83	20.70	17.24	29.41	21.94
SE	2.15	1.89	1.58	3.29	1.89
Estimated location, λ	62.04	59.92	44.76	67.20	52.46
SE	4.05	3.71	2.50	7.43	2.58
Personal discount factor, β	0.93	0.93	0.91	0.90[b]	0.90
SE		0.01	0.01		0.01
Shift in the probability of staying in the first year due to early attrition, δ	—[c]	0.05	0.08	0.07	0.08
SE		0.01	0.01	0.02	0.01
Accounting for the censoring at year 9, θ	—[c]	0.10	0.08	0.08	0.08
SE		0.01	0.01	0.02	0.01

[a] Data for Model 1 exclude cartographers; data for Model 2 include cartographers.

[b] Model estimated with fixed beta at indicated level.

[c] Model excludes the parameters that capture early attrition and censoring in year 9.

Model Fits

Figures 3.1 and 3.2 show the fit of three of the five models. Figure 3.1 shows the model fits for the two GS STEM models (Models 1 and 2 in Table 3.1). Figure 3.2 shows the model fits for the GS workforce model in which both additional parameters are included (and cartographers are included in the sample). The figures compare the observed cumulative retention rate by year of service in the data (black line) and the predicted cumulative retention rate that we simulated using the estimated model parameters (red line) and assuming the current baseline compensation system, i.e., the compensation in the absence of pay freezes and furloughs. The dotted lines are error bands associated with the cumulative retention rates in the data, as computed using the Kaplan-Meier estimator.

Overall, the models fit well, and the parameter estimates are all statistically significant from zero. Furthermore, comparisons of the simulated and actual retention profiles show that the simulated retention profiles fall within the error bands. Thus, the retention profiles predicted by the model are close to the actual data.

Figures 3.1 and 3.2 also illustrate the censoring problem at year 9, when retention falls by an unexpectedly large amount, shown by a kink in the retention profile at that year. Adding the parameter θ allowed us to retain the censored observations and account for the censoring at year 9 when we estimated the model. The model fits are excellent, regardless of whether we include or exclude the additional parameters and the censored observations. However, as shown in Table 3.1, the estimated censoring parameter θ is positive and statistically significant from zero. Censoring is estimated to shift the retention profile by between 8 percent for the GS employee subgroups (Models 3, 4, and 5) and 10 percent for the GS STEM workforce (Model 2). Thus, we used the models that include these additional parameters when we conducted our simulations, shown in Chapter Four.

Figure 3.1
Model Fits for DoD GS STEM Workers with a BA, MA, or PhD, Models 1 and 2

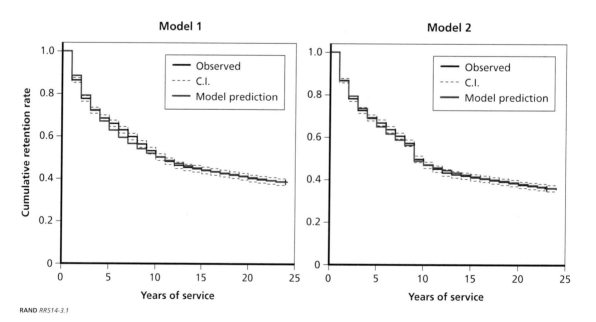

RAND RR514-3.1

Figure 3.2
Model Fit for All DoD GS Workers with a BA, MA,
or PhD

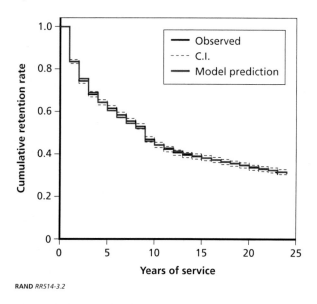

RAND *RR514-3.2*

The predictive power of the model could also be assessed by conducting out-of-sample predictions and comparing them to the actual data. We used two-thirds of the data, tracking the 1988 entry cohort through 2003 rather than through 2012, to estimate the model. We then used the estimated model to predict cumulative retention out of the sample in the eight years between 2004 and 2012. Finally, we compared the out-of-sample prediction with observed cumulative retention in the data between 2004 and 2012. Figures 3.3 and 3.4 show

Figure 3.3
Out-of-Sample Prediction and Observed Cumulative
Retention for All DoD GS Workers with an MA or PhD,
or with a BA, 1988–2003

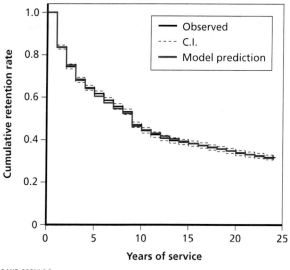

RAND *RR514-3.3*

Figure 3.4
Out-of-Sample Prediction and Observed Cumulative
Retention for STEM GS Workers with an MA or PhD,
or with a BA, 1988–2003

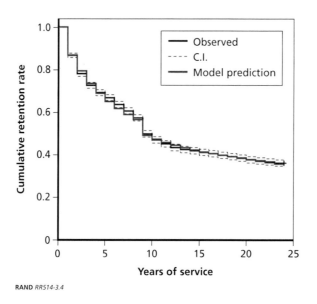

RAND *RR514-3.4*

the comparisons for the GS and STEM workforces, respectively. In both cases, the out-of-sample predictions fall within the confidence intervals of the observed cumulative retention profile. Thus, the models we estimated provide close predictions of observed retention behavior, even out-of-sample.

Parameter Estimates

The five models we estimated include two models for the STEM workforce, a simple model that excludes the extra parameters capturing early attrition and censoring in year 9 and a model that includes them. The data used to estimate Model 1 exclude cartographers. Three models are estimated for the GS workers: one with only workers with a BA, one with only workers with either an MA or PhD, and one with workers with at least a BA. These three models include the two extra parameters and also include cartographers in the data. Our preferred models are Models 2 and 5.

The estimated location parameter, λ, and the estimated mean and variance of the taste distribution, μ and σ, are scaled in thousands of dollars. Thus, the estimate of 9.14 for μ in Table 3.1 represents $9,140.

Estimated mean taste is positive in each model. It ranges from $9,140 in Model 1 to $20,660 in Model 4. A positive mean taste is consistent with federal pay being less than average private-sector pay for these subgroups. The positive taste for federal civil service offsets the lower pay of these federal workers. It is consistent with the idea that civil service workers have a positive perceived value of civil service employment over and above the monetary aspects included in the model. For example, they may value employment that offers an opportunity to serve the public or job security. In contrast, past studies estimate a negative mean taste for active and reserve service among military personnel, consistent with the fact that military pay exceeds civilian pay. The estimated mean taste in the models is the mean for a cohort of new entrants to the civil service. As the cohort progresses through its civil service career, the mean taste of those who remain will increase, not because individual tastes change over time in our model, but because those with lower taste are less likely to stay.

We found considerable heterogeneity in taste for civil service among subgroups. The estimated standard deviation of tastes ranges from $17,240 for Model 3 to $29,410 for Model 5. Interestingly, in our preferred specification for STEM workers (Model 2), the estimated standard deviation is smaller than in Model 1, $20,700. In every model, the estimated standard deviation exceeds the estimated mean of the taste distribution. A relatively large standard deviation means that controlling for compensation and shocks, there will be considerable variation in the stay/leave decisions of civil service workers at a given point in their careers.

The locality adjustments introduced in 1994 contribute to the taste variance, but in view of our estimated standard deviations, this contribution would appear to be minor. For example, in 2012, the salary for a GS 10 at step 5 ranged from approximately $57,700 to $66,200 across 85 percent of the localities, relative to an average salary of about $62,800. For a GS 13 at step 5, the respective figures were $93,400 to $103,600, with an average of $98,400. It follows that the salary range across locations is a fraction of the standard deviation of the taste.

The estimated standard deviations of the shock term are also relatively large. Since the variance of the shock term is related to λ according to the formula $\pi^2\lambda^2/6$ the standard deviation of the shock is $\pi\lambda/\sqrt{6}$. The estimated standard deviations of the shock term range from about $57,400 for Model 3 to $79,600 for Model 1. Year-to-year variation in GS salary with localities contributes to the shock standard deviation, but this variation is small (see the Appendix) compared with the estimated standard deviation of the shock. Thus, the introduction of locality adjustments has little effect on this estimate.

The relatively large estimated standard deviation of the shock term and estimated standard deviation of the taste distribution relative to the estimated mean taste are related to the fact that the estimated mean taste is positive in each model. As shown in Figures 3.1 through 3.3, over the 24 years of service in our data, about 70 percent of the employees who started in 1988 separated, and about 30 percent of entrants remained. To fit the data in which a large fraction of entrants separate by year 24, tastes must be sufficiently heterogeneous and the variance of the random shock must be large enough to induce people who have a positive mean taste for the civil service to leave. For example, the group of all GS workers with a BA, MA, or PhD has an estimated mean taste of $13,970. To induce nearly 70 percent of the entering cohort to separate by year 24, the estimated standard deviation of taste is relatively high, $21,940, as is the estimated location parameter, $52,460, which implies a $67,300 standard deviation of the shock.

The personal discount factor, β, was also estimated in our models. The estimation was successful for all but one group, GS workers with an MA or PhD. For this group, we fixed the discount factor at the estimate for all GS workers with a BA, MA, or PhD. Our estimate of β ranges from 0.90 to 0.93, with the higher rate for the GS STEM workforce. A discount factor of 0.90 means that a dollar received in one year is worth 90 cents today. These estimated discount factors are somewhat lower than the estimates for military officers and more consistent with those for enlisted personnel.

Models 2 through 5 also include a parameter to capture the shift in the probability of staying in the first year due to early attrition, δ. The drop in retention in the first year is clearly seen in Figures 3.1 through 3.3. The drop is smallest for the STEM workforce and largest for the MA and PhD subgroup of GS employees. In every model, the estimated δ is positive and significantly different from zero. The estimate is is 0.05 for STEM workers with a BA, MA, or PhD and somewhat higher—0.07 or 0.08—for all GS workers, depending on their level of education.

CHAPTER FOUR

Simulations of the Effects of Compensation Changes on Civil Service Retention

The DRM approach has been used in past studies to assess the effects of different compensation policies, such as pay raises and pay cuts; retirement reform; and targeted policies, such as separation pay and bonuses. As described in this chapter, we used the estimated models for civil service employees to assess the retention effects of a pay cut, with a focus on the three-year pay freeze of 2011–2013. We also considered the effect of an unpaid furlough that results in a one-time 3-percent decrease in annual pay. The estimated model can be used to assess many other compensation policies, though we leave that effort to future research.

Pay Freezes

Under the Federal Employee's Pay Comparability Act of 1990, civil service pay is adjusted annually to reflect annual changes in cash compensation for private-sector workers, although the President may limit the annual increase by executive order (Purcell, 2010).[1] The increase is measured as the change in the ECI in the third quarter of the previous calendar year, leading to a 15-month lag between the time at which the ECI change is measured and the effective date of the pay adjustment.

Given this methodology and the changes in the ECI published by the Bureau of Labor Statistics, the implied pay raises for GS white-collar federal workers in 2011, 2012, and 2013 should have been 0.9 percent, 1.1 percent, and 1.2 percent, respectively. Instead, because of the pay freeze, the pay raises were zero. This does not mean that federal employees could receive no pay raises. They continued to receive increases in pay associated with step increases and promotions, as indicated in the GS pay table.

We examined the effects of a three-year pay freeze in which the annual decrease in pay over the three years is 1 percent in each year. The 1-percent case closely mirrors the actual policy in 2011–2013. A 1-percent pay cut over three years translates into a 3-percent pay cut in total. We estimated the effects of the freeze for the GS STEM workforce (Model 2) and for the entire DoD GS workforce with at least a BA (Model 5).[2]

---

[1] The Pay Comparability Act stipulates that the pay raise for federal white-collar GS workers should be 0.5 percent less than the increase in the ECI for private-sector workers' wages and salaries. However, history does not bear this out. The ECI increased by 87 percent from 1990 to 2012, and GS scheduled pay including locality adjustments increased by 99 percent. Adjusting for inflation, these increases are 7 percent and 11 percent, respectively (authors' tabulations).

[2] Both of these models include the additional parameters to account for early attrition and for censoring. In the simulations described in this chapter, the censoring parameter is zero, i.e., $\theta = 0$. That is, we accounted for censoring in the data to

Our model permits us to dynamically assess the retention effects of a policy change during the transition period to the steady state, and we considered three scenarios. In the first scenario, the three-year pay freeze is permanent, i.e., Congress never restores the 3-percent cut that occurs over these three years. In this case, civil service pay is permanently 3 percent below the trajectory it would have been on in the absence of the pay freeze. Individuals in our model are forward-looking when they make their retention decisions in each year, and they recognize and, indeed, fully anticipate that their pay will never be restored.

In the second scenario, we assumed that the pay freeze is temporary, and pay is fully restored after three years. By fully restored, we mean that civil service pay is reset to its pre-freeze trajectory going forward, although we assume that back pay is not provided for the three years of lost pay raises. This would be consistent with Congress fully restoring civil service pay in 2014 to the levels that would have been in place had there been no pay freeze but without providing back pay for the lost pay raises during the three-year period. Individuals in our model anticipate that the pay cut will be fully restored at the end of the three years.

More realistically, federal employers may have some uncertainty about whether Congress will restore pay and when. If employees think that Congress will eventually restore pay some time within the next decade, we can bracket the effect of this uncertainty by considering a third scenario in which there is a ten-year delay between the start of the pay freeze and the full restoration of pay, and comparing the results with those for the case where pay is immediately restored.

While this bounding analysis does not directly model employees' uncertainty about the year in which pay will be restored, uncertainty still enters their decisionmaking process via the shock that affects the relative attractiveness of DoD civil service versus other alternatives every year. Employees anticipate these shocks, but they do not know the exact size of a shock until it occurs. Shocks will affect an individual's decision to stay or leave between the time when pay is frozen and the time when it is finally restored. We estimated sizable variances of the random shock described in Chapter Three.

We did not account for the possibility that employees revise their expectations about future pay actions and may believe that future cuts are likely. Incorporating the revision of expectations over time is a fruitful area for future research, but it is not included in the current model. Thus, our analysis of delayed restoration should be viewed as a first look at this issue.

As will be shown, the retention effect of the pay freeze depends on whether pay is later restored and how long it takes before that restoration occurs. The effects differ in the short term and the long term, and they differ for junior, mid-career, and senior personnel.

A Three-Year Pay Freeze with No Future Restoration of Pay
Figure 4.1 shows the steady-state effects on DoD GS and GS STEM retention of a three-year pay freeze of 1 percent per year, after which pay is never restored. By steady state, we mean the effects of the pay freeze on retention for a member who spends his or her entire career under the lower pay. Assuming a 40-year career, it would take 40 years for federal employees serving at the time of the policy to completely flow out of the civil service and the new steady state to be achieved. Thus, Figure 4.1 shows the effects of the policy 40 years after the pay freeze began.

estimate the model, but our simulations of workforce retention under different compensation policies assumed that censoring is no longer a factor affecting the probability of staying in the civil service.

Figure 4.1
Steady-State Effects of a Permanent Three-Year, 1-Percent Annual Pay Freeze on GS Workforce Retention and GS STEM Workforce Retention

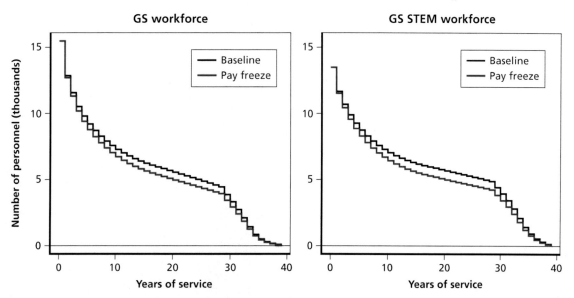

NOTE: The graphs show model steady-state simulations of the retention effects of a three-year pay freeze that is permanent in the sense that pay is never restored to its pre-freeze trajectory. The right-hand graph shows the effects for GS STEM workers with at least a BA degree, and the left-hand graph shows the effects for GS workers with at least a BA degree. The graphs show the effects of a 1-percent pay freeze for each of the three years.
RAND RR514-4.1

Figure 4.2 shows selected years during the transition to the steady state for the entire GS workforce with a BA or higher degree, and Figure 4.3 shows the dynamics of the transition for the subset of this workforce that are in the STEM occupations. The figures show the size of the force retained by years of service after 3 years following the enactment of the pay freeze, after 5 years, after 10 years, and after 20 years.

Because pay is not restored, the steady-state results in Figure 4.1 can be thought of as the long-term effects of a permanent 3-percent pay cut. Given our model estimates, the 1-percent per year pay freeze over three years reduces the better-educated GS workforce that is retained by 7.3 percent in the steady state and reduces the retained GS STEM workforce by 8.5 percent.

The retention responses to the pay freeze are larger among workers in the earlier career years than among retirement-eligible personnel. The effect of the pay cut changes once individuals are eligible for immediate retirement benefits, e.g., after 30 years of service. For these individuals, the pay cut affects both the value of staying and the value of leaving. Staying another year in the civil service means trading off the higher benefits associated with another year against the forgone benefits that could have been realized by leaving immediately. Thus, the pay cut affects both the immediate wages of retirement-eligible employees and their immediate cash benefits if they were to leave, so the relative difference between civil service and external pay is mitigated. Consequently, retirement-eligible personnel are less sensitive to the pay cut than those with fewer years of service.

The pay cut affects both cash compensation and the value of future retirement benefits, given that all three elements of FERS—the defined benefit plan, the defined contribution plan, and Social Security benefits—depend on civil service pay. Given the estimated discount factors

Figure 4.2
**Transition Effects of a Permanent Three-Year, 1-Percent Annual Pay Freeze on GS Workforce
Retention 3, 5, 10, and 20 Calendar Years After the Pay Freeze Began**

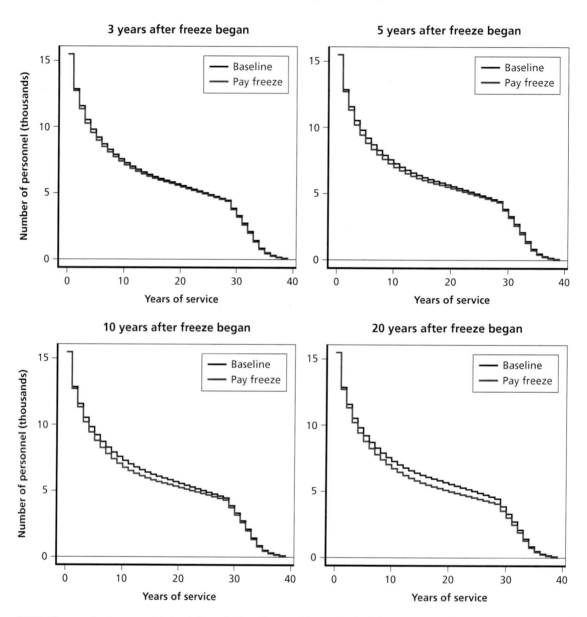

NOTE: The graphs show model simulations during the transition period of the retention effects of a three-year
pay freeze that is permanent in the sense that pay is never restored to its pre-freeze trajectory. The graphs are
for all GS workers with at least a BA degree and show the effects of a 1-percent pay freeze for each of the three
years. The upper left-hand graph shows the effects at *t*+3, or three years after the pay freeze began; the upper
right-hand graph shows the effects at *t*+5; the lower left-hand graph shows the effects at *t*+10; the lower
right-hand graph shows the effects at *t*+20.
RAND RR514-4.2

of 0.90 for GS workers and 0.93 for GS STEM workers, junior personnel significantly discount
benefits that occur far out in the future. That is, the effect of the pay cut on expected retirement
benefits is small for junior personnel but increases as individuals approach retirement-eligibil-
ity age and years of service when retirement benefits can begin. For example, a member with
30 years of service can retire with immediate benefits at the minimum retirement age (assumed

to be 56 in our model). As civil service employees approach 30 years of service, they discount future retirement benefits by less and less, so the effect of the pay cut on expected future retirement benefits gets larger. Still, the decrease in retention is smaller after many years of service than at early career years, for the reasons discussed above.

The effects in the transition to the steady state are shown for selected years in Figure 4.2 for the GS workforce and Figure 4.3 for the GS STEM workforce. The key finding is that the

Figure 4.3
Transition Effects on GS STEM Workforce Retention of a Permanent Three-Year, 1-Percent Annual Pay Freeze 3, 5, 10, and 20 Calendar Years After the Pay Freeze Began

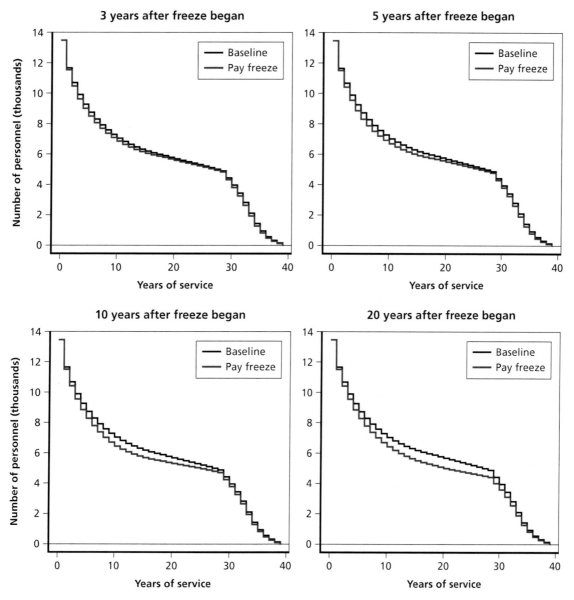

NOTE: The graphs show model simulations during the transition period of the retention effects of a three-year pay freeze that is permanent in the sense that pay is never restored to its pre-freeze trajectory. The graphs are for GS STEM workers with at least a BA degree and show the effects of a 1-percent pay freeze for each of the three years. The upper left-hand graph shows the effects at *t*+3, or three years after the pay freeze began; the upper right-hand graph shows the effects at *t*+5; the lower left-hand graph shows the effects at *t*+10; the lower right-hand graph shows the effects at *t*+20.

RAND *RR514-4.3*

retention effects of the three-year pay freeze without restoration build up over time. By the end of the third year, corresponding to the end of the freeze, the GS force size that is retained has fallen by 2.2 percent, and the retained GS STEM workforce has decreased by 2.5 percent. That is, only 2.2 percent of the total steady-state 7.3-percent drop in the size of the retained GS workforce and only 2.5 percent of the 8.5-percent drop in the GS STEM workforce occurred by the third year. By year 5, about 45 percent of the total steady-state drop in the retained workforce has occurred—3.3 percent for the GS workforce and 3.7 percent for the GS STEM workforce. Even by year 20, the full steady-state effect has not been reached.

The retention effect of a permanent pay cut unfolds slowly because of two factors. First, those at the middle and end of their civil service careers at the time of the pay freeze have fewer remaining years with lower pay relative to those who are at the beginning of their careers. The permanent pay cut affects future cash compensation as well as future retirement benefits, but those with more years of service have relatively few years with lower pay, and the impact on their expected future retirement benefits will be correspondingly less. Second, as discussed above in the context of the steady state, the decrease in retention as a result of the pay freeze is initially largest among those in their early career years, and the decline in retention grows larger over time for those with more years of service as these individuals gain experience and progress through their careers.

A Three-Year Pay Freeze with Immediate Restoration of Pay

In this subsection, we present results of a scenario in which Congress restores pay immediately following the pay freeze to the pre-freeze trajectory going forward, though it does not provide back pay for the three years of lost raises.

The main result is that the pay freeze has virtually no effect on retention in either the steady state or during the transition period. While the pay freeze reduces the value of staying because of the three years of lost back pay, civil service employees are forward-looking and recognize that civil service pay will be reset to the pre-freeze trajectory immediately after the freeze ends. The restoration of pay offsets the negative effects of the pay freeze on the value of staying, despite the lack of back pay, so retention remains essentially unchanged.

Figure 4.4 shows the steady-state results for the three-year, 1-percent per year pay freeze after which pay is restored for the GS and GS STEM workforces. In both cases, the size of the steady-state workforce that is retained is unchanged, because in the new steady state, all workers are under the new policy of fully restored pay. Figure 4.5 shows the dynamics during the transition period for the GS workforce, when individuals anticipate the restoration of pay and their retention behavior is unchanged. We do not show the results for the GS STEM workforce, but they are identical to those shown in Figure 4.5. That is, civil service employees expect the restoration of their pay, and although back pay is not provided, their retention behavior is virtually unchanged during the transition period.

A Three-Year Pay Freeze with Full but Delayed Restoration of Pay

In our third scenario, Congress restores pay to the pre-freeze levels, but with a ten-year delay. Thus, a currently serving civil service worker must wait 13 years from the time the three-year pay freeze begins before pay is restored. We could consider other delay periods, and indeed the case of the permanent pay cut is the extreme case in which the delay is 40 years. However, the main points are illustrated with the assumption of a ten-year delay.

Figure 4.4
Steady-State Effects of a Three-Year, 1-Percent Annual Pay Freeze with Full and Immediate Restoration on GS Workforce and GS STEM Workforce Retention

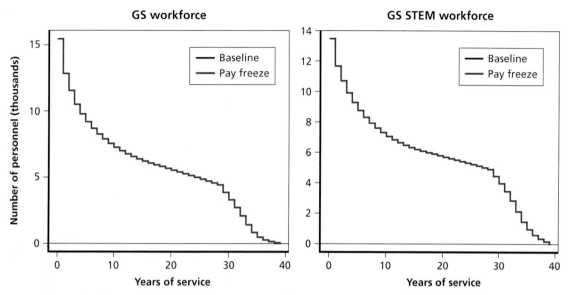

NOTE: The graphs show model steady state simulations of the retention effects of a three-year pay freeze that is immediately restored in the sense that pay is reset to its pre-freeze trajectory at *t*+4, or the year immediately after the three-year pay freeze ends. The left-hand graph shows the effects for GS workers with at least a BA degree, and the right-hand graph shows the effects for GS STEM workers with at least a BA degree. The panels show the effects of a 1-percent pay freeze for each of the three years.

RAND *RR514-4.4*

There is no change in steady-state retention as a result of the pay freeze, even when the restoration of pay is delayed for ten years and no back pay is provided, as shown in Figure 4.6. The steady state occurs when all new entrants are under the restored pay regime, so their retention behavior is unchanged relative to the base case.

However, as Figure 4.7 makes clear for the GS workforce and Figure 4.8 shows for the GS STEM workforce, retention changes during the transition to the steady state. When restoration is delayed, retention falls in the initial years by almost the same amount that it falls in the initial years when pay is not restored. For example, after five years, the retained GS workforce decreases by 3.3 percent when pay is not restored but by 2.8 percent when pay is restored. However, the drop in retention slows down in the transition period when pay is eventually restored. In year 20, the drop in the retained GS workforce is 1.4 percent when pay is eventually restored but 6.8 percent when it is not. Similar qualitative results are found for the GS STEM workforce, as shown in Figure 4.8. Retention falls in the transition period when the restoration of the pay freeze is delayed, but eventually the drop slows down and reverses.

The workforce dynamics shown in Figures 4.7 and 4.8 are not surprising. In the initial years, civil service employees are exposed to the lower pay associated with the pay freeze until pay is restored to the pre-freeze level, and they discount the future restoration of pay. As time elapses, their exposure is less, and they discount the future restoration of pay less. Eventually, employees exposed to the lower pay flow out of the civil service. All employees are then under the restored pay system, resulting in the new steady state.

Comparison of the results in Figures 4.5 and 4.6 with those in Figures 4.7 and 4.8 shows that delays have real effects on retention. When there is no delay, the pay freeze has virtually no

Figure 4.5
Transition Effects of a Three-Year, 1-Percent Annual Pay Freeze with Full and Immediate Restoration on GS Workforce Retention 3, 5, 10, and 20 Calendar Years After the Pay Freeze Began

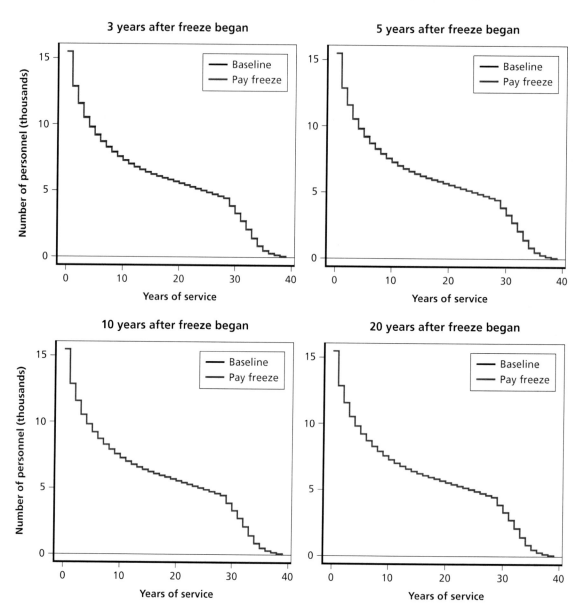

NOTE: The graphs show model simulations during the transition period of the retention effects of a three-year pay freeze that is immediately restored in the sense that pay is reset to its pre-freeze trajectory at *t*+4, or the year immediately after the three-year pay freeze ends. The graphs are for GS workers with at least a BA degree and show the effects of a 1-percent pay freeze for each of the three years. The upper left-hand graph shows the effects at *t*+3, or three years after the pay freeze began; the upper right-hand graph shows the effects at *t*+5; the lower left-hand graph shows the effects at *t*+10; the lower right-hand graph shows the effects at *t*+20.
RAND *RR514-4.5*

Figure 4.6
Steady-State Effects of a Three-Year, 1-Percent Annual Pay Freeze with Full and Ten-Year Delayed Restoration on GS Workforce and GS STEM Workforce Retention

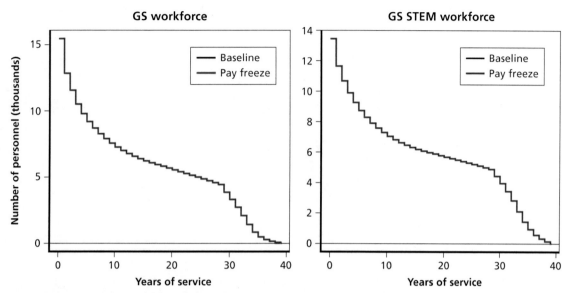

NOTE: The graphs show model steady-state simulations of the retention effects of a three-year pay freeze that is restored with a ten-year delay in the sense that pay is reset to its pre-freeze trajectory at *t*+14, or ten years after the three-year pay freeze ends. The left-hand graph shows the effects for GS workers with at least a BA degree, and the right-hand graph shows the effects for GS STEM workers with at least a BA degree. The graphs show the effects of a 1-percent pay freeze for each of the three years.
RAND *RR514-4.6*

retention effects during the transition period. But the delays cause a reduction in retention for some amount of time, and the longer the delay, the longer the time that retention is reduced.

Unpaid Furloughs

In 2013, federal civil service employees experienced unpaid furloughs as a result of sequestration. Initially, in March 2013, DoD civil service employees faced the possibility of a 22-day furlough, but when furloughs began, the length of time was reduced to 11 days. In August 2013, the time was further reduced to six days. A six-day unpaid furlough corresponds to a 2.3-percent cut in pay (assuming 261 non-weekend workdays in a year), or about 3 percent. The cut in pay does not affect the computation of retirement benefits, because benefits are based on the individual's pay rate. Federal employees also had unpaid furloughs as a result of the partial government shutdown in October 2013, but Congress voted to restore that lost pay.

We used the DRM estimates to simulate the effect of an unpaid six-day furlough by considering the effect of a 3-percent cut in annual pay in one year. The analysis did not account for any value associated with leisure or nonmarket activities during the furlough. Insofar as employees place some value on having more leisure time, the value of the loss to workers from the unpaid furlough is less than 3 percent. Therefore, we overstate any negative effect on retention in that case. The analysis also did not account for any readjustment of expectations about additional future furloughs (and resulting unpaid work), so we may possibly understate any negative effect on retention.

Figure 4.7
Transition Effects on GS Workforce Retention of a Three-Year, 1-Percent Annual Pay Freeze with Full and Ten-Year Delayed Restoration 3, 5, 10, and 20 Calendar Years After the Pay Freeze Began

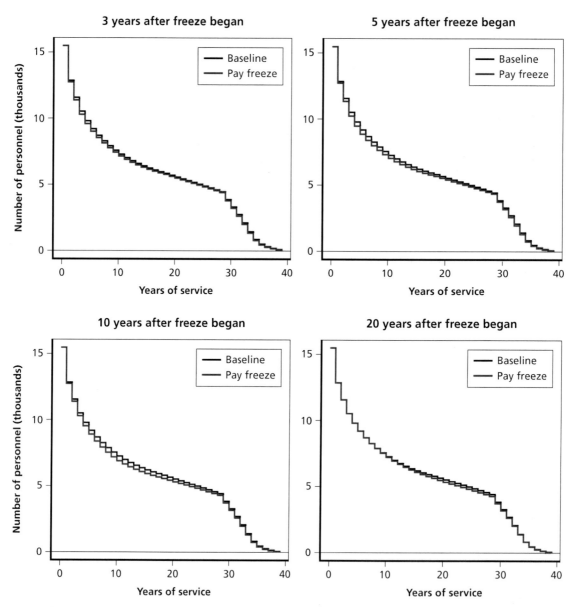

NOTE: The graphs show model simulations during the transition period of the retention effects of a three-year pay freeze that is restored with a ten-year delay in the sense that pay is reset to its pre-freeze trajectory at $t+14$, or ten years after the three-year pay freeze ends. The graphs are for GS workers with at least a BA degree and show the effects of a 1-percent pay freeze for each of the three years. The upper left-hand graph shows the effects at $t+3$, or three years after the pay freeze began; the upper right-hand graph shows the effects at $t+5$; the lower left-hand graph shows the effects at $t+10$; the lower right-hand graph shows the effects at $t+20$.
RAND RR514-4.7

Figure 4.8
Transition Effects on GS STEM Workforce Retention of a Three-Year, 1-Percent Annual Pay Freeze with Full and Ten-Year Delayed Restoration 3, 5, 10, and 20 Calendar Years After the Pay Freeze Began

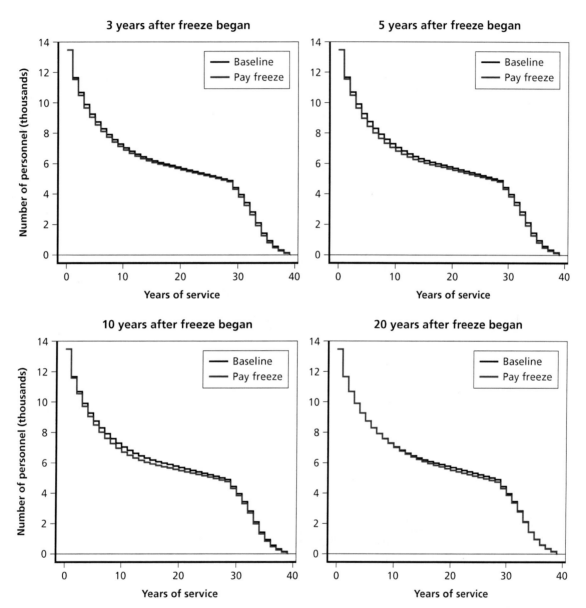

NOTE: The graphs show model simulations during the transition period of the retention effects of a three-year pay freeze that is restored with a ten-year delay in the sense that pay is reset to its pre-freeze trajectory at $t+14$, or ten years after the three-year pay freeze ends. The graphs are for GS STEM workers with at least a BA degree and show the effects of a 1-percent pay freeze for each of the three years. The upper left-hand graph shows the effects at $t+3$, or three years after the pay freeze began; the upper right-hand graph shows the effects at $t+5$; the lower left-hand graph shows the effects at $t+10$; the lower right-hand graph shows the effects at $t+20$.
RAND RR514-4.8

Figure 4.9 shows the dynamics of change for the retained GS workforce when the 3-percent pay cut lasts for only one year, as was the case during the sequestration in August 2013. Figure 4.10 shows the results for the GS STEM workforce.

Figure 4.9
Transition Effects on GS Workforce Retention of an Unpaid Furlough Resulting in a 3-Percent One-Time Cut in Annual Pay

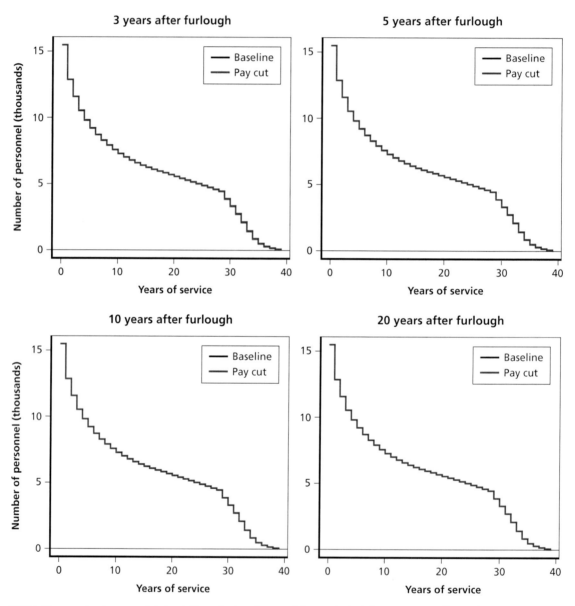

NOTE: The graphs show model simulations during the transition period of the retention effects of a one-time 3-percent cut in annual pay that is restored to the pre-cut levels in the following year. The graphs are for GS workers with at least a BA degree and show the effects of a 3-percent pay cut. The upper left-hand graph shows the effects at *t*+3, or three years after the furlough; the upper right-hand graph shows the effects at *t*+5; the lower left-hand graph shows the effects at *t*+10; the lower right-hand graph shows the effects at *t*+20.
RAND *RR514-4.9*

Figure 4.10
Transition Effects on GS STEM Workforce Retention of an Unpaid Furlough Resulting in a 3-Percent One-Time Cut in Annual Pay

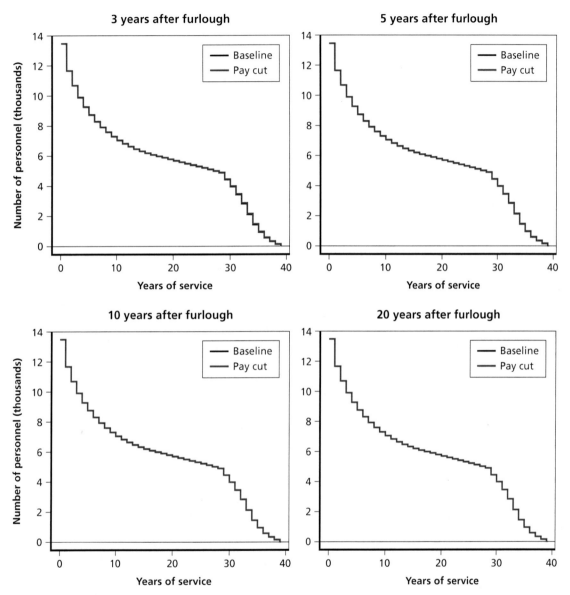

NOTE: The graphs show model simulations during the transition period of the retention effects of a one-time 3-percent cut in annual pay that is restored to the pre-cut levels in the following year. The graphs are for GS STEM workers with at least a BA degree and show the effects of a 3-percent pay cut. The upper left-hand graph shows the effects at t+3, or three years after the furlough; the upper right-hand graph shows the effects at t+5; the lower left-hand graph shows the effects at t+10; the lower right-hand graph shows the effects at t+20.
RAND RR514-4.10

The unpaid six-day furlough had no discernible effect on retention for either the GS or GS STEM workforce. Figures 4.9 and 4.10 show the effect on retained workforce sizes after 1, 3, 5, and 10 years for each workforce. The drop in the retained force size is minimal. The effect is tiny because the pay cut lasts only one year and pay reverts back to normal the following year. While a few federal employees are induced to leave, most anticipate that pay will go back to normal and therefore do not change their retention behavior. Of course, if unpaid furloughs became "the new normal" and individuals expect more furloughs in the future, the effects would last longer and be larger. On the other hand, the small retention effect we found could be even smaller, depending on how workers value the leisure associated with the furlough.

CHAPTER FIVE

Policy Implications and Areas for Future Research

We simulated the responsiveness of DoD civil service employee retention to changes in compensation using a logically consistent, empirically based methodology and found that a three-year, 1-percent per year pay freeze that was not later restored would decrease the GS STEM workforce and the broader GS workforce of better-educated workers. We also found that the effects on retention in the first few years after the pay freeze were largest among employees in their early career years and smaller for the most junior personnel and for those nearing 30 years of service.

The simulation results show the supply response to compensation changes but do not indicate that response relative to requirements. Put differently, our simulations were performed under the assumption that current pay levels are adequate but not excessive. Under this assumption, the effect of the pay freeze is a bit larger for the GS STEM workforce than for the broader GS workforce. Our analysis did not include the entire STEM workforce in DoD, however; the larger STEM workforce includes personnel under pay plans other than the GS system.

The simulations provide an estimate of the extent to which a pay freeze can adversely affect the size of the civil service workforce that is retained, but the this does not necessarily imply that federal pay should therefore be increased or restored. Assessments of the adequacy of federal pay levels must consider current and future workforce requirements and the current and future supply of personnel to meet those requirements. Insofar as planners deem that the supply is inadequate relative to the requirements, the DRM capability can provide empirically based simulations of the impact on retention of alternative compensation policies aimed to increase supply. While this analysis does not include cost estimates of alternative policies, we have made such estimates in related analyses for military personnel, and that capability could be extended to the civil service.

Our analyses could be extended in many ways, providing a fruitful area for future research. The DRM capability can be used to simulate the retention effects of other policies of interest. For example, Katz (2013) reports that lawmakers in Congress, as well as the President, have been considering a number of proposals to change federal compensation. These include increasing federal pension contributions, eliminating the FERS annuity supplement, lowering the FERS basic plan multiplier, and using the so-called chained Consumer Price Index to calculate cost-of-living adjustments for federal retirees. Other policies include incentive pays targeted toward the retention of key subgroups of personnel and severance pay, early retirement, and other drawdown policies to induce voluntary separations. The retention effects of all of these proposals can be assessed with the DRM capability, with appropriate adjustment and updating of our simulation code.

The DRM capability could also be extended to other occupational areas within DoD, including the cyber workforce; to other pay systems such as the STEM workforces in the various demonstration programs and the Wage Grade workforce; to specific demographic groups, such as women and minorities; and to specific locations of interest, such as Hawaii. Furthermore, with appropriate data, the DRM capability could be applied to civil service workforces in other agencies within the federal government, including the Department of Veterans Affairs, the Department of Homeland Security, and the various agencies that make up the intelligence community.

The DRM model itself could be extended in important ways in future research. First, we could incorporate into the model the fact that some civil service personnel leave and then return. Returnee behavior has already been introduced in models of reserve-component participation in the military, and it can be introduced in the civil service model as well. Second, we could incorporate changes in expectations about future policy changes. Pay freezes and furloughs can lead to changed expectations about the likelihood of future pay freezes and furloughs. Workers may view future pay as more uncertain. Such an extension of the model would require incorporating a model of how workers form and change their expectations.

Third, the model could be extended to create a "total force" model of DoD workforce dynamics and the effects of compensation on those dynamics where Total Force includes active and reserve military personnel and DoD civilians (but not contractors). As shown in Gates et al. (2008), about half of new DoD civil service hires have prior military service. Furthermore, many DoD civilians also participate in the reserve components as drilling selected reservists. Consequently, changes in compensation and personnel policy in either the active component, the reserve component, or the federal civil service are likely to affect retention in all three parts of the DoD workforce in interrelated ways. As an example of this, Mattock, Asch, and Hosek (2012) find that increases in reserve-component retirement benefits reduce mid-career retention in the active component. RAND has already developed a unified DRM capability to provide logically consistent and empirically based estimates of the effects of compensation policy on active-component retention and reserve-component participation; this capability could be extended to include DoD civil service employment.

Given the size of the federal workforce and, in the context of DoD, its contribution to military readiness, it is critical that planners and policymakers have the capability to understand how changes in compensation and personnel policy affect that workforce. The DRM can provide such a capability, and the analysis presented in this report represents a step toward it.

Locality Adjustments in General Schedule Salaries

Recognizing geographical disparities in the wage structure of private-sector workers who are comparable in terms of occupation and work level to those employed by the federal government, federal officials in 1994 introduced locality adjustments to GS salaries. The annual GS salary adjustment has since then been based on adjustment to the base GS salary table plus a locality-specific adjustment, both stated as a percentage increase of the then-current salary. In this appendix, we describe the annual across-the-board and locality adjustments to GS salaries, the cumulative percentage increase in the locality adjustment since 1993 by locality, and a simple method for approximating the total percentage increase in GS pay for a locality since 1993.

The National Compensation Survey (NCS), administered by the Bureau of Labor Statistics, is used to set and adjust GS salaries, including annual across-the-board pay adjustments to the base GS schedule and locality adjustments. The across-the-board pay adjustments are based on the annual change in the ECI "for wages and salaries for private industry workers less 0.5 percentage points, as of September, 15 months preceding the January adjustment."[1] The ECI is a national measure of employment cost. Locality pay adjustments began in 1994 for 33 areas and the "rest of U.S." and had expanded to 41 areas by 2013. The 33 areas are metropolitan statistical areas (MSAs) established by the Office of Management and Budget and augmented to include combined statistical areas (CSAs). CSAs are "aggregates of adjacent metropolitan or micropolitan statistical areas that are linked by commuting ties."[2] The 33 MSA/CSAs are a subset of all U.S. metropolitan areas, which numbered 381 in 2013. Locality pay adjustments are based on locality pay surveys, which are components of the NCS. The NCS collects wage data from nonfederal establishments, using "probability sampling methods to choose jobs from which to collect the wage and benefit data." The selected jobs are matched to GS occupational series on the basis of both occupation and work level; law requires the inclusion of work level in the matching. Only the NCS acquires work-level detail.[3]

Table A.1 shows the annual percentage adjustments. The first column, GS base, shows the annual percentage increase in the GS base salary table. Until 1994, this was the single, nationwide table. In the locality adjustments introduced in 1994, all of the 4-percent increase in GS salary was allocated to establishing differentials by locality. Since then, however, about one-fourth of the total annual adjustment has gone to locality adjustments. The second column shows the locality adjustment percentage, which should be interpreted as the percentage of

---

[1] U.S. Office of Personnel Management, undated c.

[2] U.S. Census Bureau, rev. 2013.

[3] U.S. Office of Personnel Management, undated c.

Table A.1
Percentage Adjustments to GS Salaries, 1990–2012

Year	GS Base, Annual Percentage (1)	Locality Adjustment, Percentage (2)	Total Percentage Change (3)	Cumulative Base (1993 = 100) (4)	Cumulative Total (1993 = 100) (5)	CPI-U Index (1993 = 100) (6)	Inflation-Adjusted index (7)
1990	3.6		3.6	88.6	88.9	87.9	1.01
1991	4.1		4.1	91.9	92.5	92.9	1.00
1992	4.2		4.2	95.8	96.4	97.0	0.99
1993	3.7		3.7	100.0	100.0	100.0	1.00
1994	0.0	4.0	4.0	100.0	104.0	102.6	1.01
1995	2.0	0.6	2.6	102.0	106.7	105.5	1.01
1996	2.0	0.4	2.4	104.0	109.3	108.6	1.01
1997	2.3	0.7	3.0	106.4	112.5	111.1	1.01
1998	2.3	0.6	2.9	108.9	115.8	112.9	1.03
1999	3.1	0.5	3.6	112.3	120.0	115.4	1.04
2000	3.8	1.0	4.8	116.5	125.7	119.3	1.05
2001	2.7	1.0	3.7	119.7	130.4	122.7	1.06
2002	3.6	1.0	4.6	124.0	136.4	124.6	1.09
2003	3.1	1.0	4.1	127.8	142.0	127.5	1.11
2004	2.7	1.4	4.1	131.3	147.8	130.9	1.13
2005	2.5	1.0	3.5	134.6	153.0	135.4	1.13
2006	2.1	1.0	3.1	137.4	157.7	139.7	1.13
2007	1.7	0.5	2.2	139.7	161.2	143.6	1.12
2008	2.5	1.0	3.5	143.2	166.8	149.1	1.12
2009	2.9	1.0	3.9	147.4	173.3	148.5	1.17
2010	1.5	0.5	2.0	149.6	176.8	150.9	1.17
2011	0.0	0.0	0.0	149.6	176.8	155.7	1.14
2012	0.0	0.0	0.0	149.6	176.8	159.0	1.11

NOTE: This table is an extract from Table 1 in Purcell (2008).

the pay increase that is allocated to locality adjustments. The third column, total percentage change, is the sum of the first and second columns. The fourth and fifth columns show the cumulative increase in the base GS salary and the base GS salary plus the locality adjustment since 1990 in nominal terms. The sixth column has the consumer price index–urban, which we use to adjust for inflation, and the seventh column has the inflation-adjusted cumulative total salary with 1993 chosen as the base year (1993 = 100).

Table A.2 shows the cumulative locality pay adjustment by locality for 1994, 2002, and 2010, relative to 1993. These years are representative of locality adjustments; the omitted years show a similar pattern. The similarity of the pattern over time reflects the stability, or persistence, in the evolution of local wage structures—localities with a higher locality increase in 1994 also tended to have a higher cumulative increase to 2010. Another, although indirect, reflection of this stability is the coefficient of variation (the ratio of the standard deviation to the average), which is relevant here because the average percentage increase to 2002 is higher than the first-year increase, and the average percentage increase to 2010 is still higher. The

Table A.2
Cumulative Locality Adjustments to GS Salaries from 1993 to 1994, 2002, and 2010

MSA/CSA	1994	2002	2010
Atlanta, GA	3.86	9.74	19.29
Boston-Worcester-Lawrence, MA-NH-ME-CT	5.47	13.57	24.80
Buffalo-Niagara-Cattaraugus, NY			16.98
Chicago-Gary-Kenosha, IL-IN-WI	5.34	14.58	25.10
Cincinnati-Hamilton, OH-KY-IN	4.22	12.09	18.55
Cleveland-Akron, OH	3.34	10.33	18.68
Columbus, OH		10.70	17.16
Dallas-Fort Worth, TX	4.21	10.90	20.67
Dayton-Springfield, OH	3.77	9.62	16.24
Denver-Boulder-Greeley, CO	4.54	13.34	22.52
Detroit-Ann Arbor-Flint, MI	4.84	14.71	24.09
Hartford, CT		14.11	25.82
Houston-Galveston-Brazoria, TX	6.52	18.61	28.71
Huntsville, AL	4.10	9.08	16.02
Indianapolis, IN	3.68	8.85	14.68
Kansas City, MO-KS	3.30	9.28	
Los Angeles-Riverside-Orange County, CA	5.69	16.05	27.16
Memphis, TN-AR-MS	3.09		
Miami-Fort Lauderdale, FL		12.45	20.79
Milwaukee-Racine, WI		10.05	18.10
Minneapolis-St. Paul, MN-WI		11.56	20.96
New York-Northern New Jersey-Long Island, NY-NJ-CT-PA	5.77	15.23	28.72
Norfolk-Virginia Beach-Newport News, VA-NC	3.28		
Oklahoma City, OK	3.34		
Orlando, FL		8.67	
Philadelphia-Wilmington-Atlantic City, PA-NJ-DE-MD	4.96	12.11	21.79
Phoenix-Mesa-Scottsdale, AZ			16.76
Pittsburgh, PA		9.52	16.37
Portland-Salem, OR-WA		11.64	20.35
Raleigh-Durham-Cary, NC			17.64
Richmond-Petersburg, VA		9.67	16.47
Sacramento-Yolo, CA	3.69	11.99	22.20
San Diego-Carlsbad-San Marcos, CA			24.19
St. Louis, Missouri-IL	3.09	8.98	
Salt Lake City-Ogden, UT	3.09		
San Antonio, TX	3.09		
San Diego, CA	3.88	12.70	
San Francisco-Oakland-San Jose, CA	6.18	19.04	35.15
Seattle-Tacoma- Bremerton, WA	3.92	11.77	21.81
Washington-Baltimore, District of Columbia-MD- VA-WV	4.23	11.48	24.22
Rest of United States	3.09	8.64	14.16
Interim geographic adjustments	8.00		
GS base increase	4.00	3.60	1.50
Standard deviation	1.02	2.74	4.77
Average	4.20	11.91	21.13
Coefficient of variation	0.24	0.23	0.23

NOTE: Drawn from tables for 1994, 2002, and 2010 in U.S. Office of Personnel Management, undated b.

coefficient of variation "normalizes" the comparison of standard deviations by making them relative to the average increase. As seen in the last row of Table A.2, the coefficient of variation is almost exactly the same in 1994, 2002, and 2010, suggesting that the salary differentiation established in the first year of locality adjustments set the pattern for locality differentials, and the variation across the localities has remained largely the same since then. Visually comparing the percentage increase in Table A.2 offers a more direct comparison. Further, the correlation between the entries in both 1994 and 2002 is 0.92, that in both 1994 and 2010 is 0.87, and that in both 2002 and 2010 is 0.94. This persistence suggests that when we estimate our model using civil service wage data for 2011, the locality differences in effect enter the model as fixed effects that contribute to the variation in what we refer to as "taste."

The tables provided by the U.S. Office of Personnel Management show base GS salary by year and locality adjustments by locality; the latter are stated as a cumulative percentage increase since 1993 and as a percentage increase over the previous year (U.S. Office of Personnel Management, undated b). However, there is no table showing the year-by-year locality adjustments for each locality, although this information is necessary for a precise calculation of a locality's cumulative total increase (base plus locality).

It is possible to approximate a locality's cumulative total increase with available data as follows. Consider a two-year span following a base year. Taking the base year salary as 1 and letting α_1 and α_2 be the relative increases in *base* GS salary in years 1 and 2, and b_{1i} and b_{2i} be the relative locality increases for locality i in years 1 and 2, the cumulative total increase over two years is the product $(1 + a_1 + b_{1i})(1 + a_2 + b_{2i})$. The natural log of this product is $\ln(1 + a_1 + b_{1i}) + \ln(1 + a_2 + b_{2i}) \approx (a_1 + a_2) + (b_{1i} + b_{2i})$. The cumulative relative increase is approximately $\exp((a_1 + a_2) + (b_{1i} + b_{2i}))$. Applying this result, the cumulative relative increase since 1993 for a locality is approximately equal to e, an exponent equal to the sum of the cumulative percentage increase in base GS salary and the cumulative percentage increase in the locality adjustment, minus 1. These quantities are shown in Tables A.1 and A.2. For instance, the cumulative percentage increase in the base GS salary (for any given grade and step) from 1993 to 2010 is 0.496, the cumulative locality adjustment for Atlanta is 0.193, and the cost-of-living increase is 0.509, giving an exponent of (0.496 + 0.193 − 0.509) = 0.180. The cumulative percentage increase in total GS salary for Atlanta is then approximately $100 * (e^{0.180} - 1) = 19.7$, or about 20 percent.

Using this approach, we computed the approximate cumulative increase in total GS salary for cumulative locality adjustments that cover the range of those shown in Table A.2 for 2010. The results are given in Table A.3. As seen, the cumulative increase in total GS salary is quite close to the cumulative locality adjustment. This results because the cumulative increase in base GS salary, 0.496, is virtually equal to the cumulative increase in inflation, 0.509, so these values offset one another and leave only the locality adjustment to drive the cumulative increase in total GS salary.

Table A.3
**Approximate Cumulative Increase in Total GS
Salaries from 1993 to 2010 Across the Range
of Cumulative Locality Adjustments**

Locality Adjustment from 1993 to 2010 (percentage)	Approximate Increase in Total GS Salary from 1993 to 2010 (percentage)
14	14
16	16
18	18
20	21
22	23
24	25
26	28
28	31
30	33
32	36
34	39
36	41

References

Aguirregabiria, Victor, and Pedro Mira, "Dynamic Discrete Choice Structural Models: A Survey," *Journal of Econometrics*, Vol. 156, No. 1, 2010, pp. 38–67.

Asch, Beth J., *The Pay, Promotion, and Retention of High-Quality Civil Service Workers in the Department of Defense*, Santa Monica, Calif.: RAND Corporation, MR-1193-OSD, 2001. As of May 16, 2014: http://www.rand.org/pubs/monograph_reports/MR1193.html

Asch, Beth, Steven Haider, and Julie Zissimopoulos, "The Effects of Workforce-Shaping Tools on Retirement: The Case of Department of Defense Civil Service," *Journal of Public Health Management Practices*, November 2009 (Supplement), pp. S64–S72.

———, "Financial Incentives and Retirement: Evidence from Federal Civil Service Workers," *Journal of Public Economics*, Vol. 89, 2005, pp. 427–440.

Asch, Beth J., James Hosek, and Michael G. Mattock, *A Policy Analysis of Reserve Retirement Reform*, Santa Monica, Calif.: RAND Corporation, MG-378-OSD, 2013. As of May 15, 2014: http://www.rand.org/pubs/monographs/MG378.html

Asch, Beth J., James Hosek, Michael G. Mattock, and Christina Panis, *Assessing Compensation Reform: Research in Support of the 10th Quadrennial Review of Military Compensation*, Santa Monica, Calif.: RAND Corporation, MG-764-OSD, 2008. As of May 15, 2014: http://www.rand.org/pubs/monographs/MG764.html

Asch, Beth J., and John T. Warner, *Separation and Retirement Incentives in the Federal Civil Service: A Comparison of the Federal Employees Retirement System and Civil Service Retirement System*, Santa Monica, Calif.: RAND Corporation, MR-986-OSD, 1999. As of May 15, 2014: http://www.rand.org/pubs/monograph_reports/MR986.html

Bajari, Patrick, C. Lanier Benkard, and Jonathan Levin, "Estimating Dynamic Models of Imperfect Competition," *Econometrica*, Vol. 75, No. 5, September 2007, pp. 1331–1370.

Biggs, Andrew G., and Jason Richwine, "Comparing Federal and Private Sector Compensation," Washington, D.C.: American Enterprise Institute for Public Policy Research, Working Paper 2011-02, 2011. As of October 7, 2013: http://www.aei.org/papers/economics/fiscal-policy/labor/comparing-federal-and-private-sector-compensation/

Black, Matthew, Robert Moffitt, and John Warner, "The Dynamics of Job Separation: The Case of Federal Employees," *Journal of Applied Econometrics*, Vol. 5, No. 3, 1990a, pp. 245–262.

———, "Reply to Comment by Glenn Gotz on 'The Dynamics of Job Separation: The Case of Federal Employees'," *Journal of Applied Econometrics*, Vol. 5, No. 3, 1990b, pp. 269–272.

Borkovsky, Ron, Ulrich Doraszelski, and Yaroslav Kryukov, "A Dynamic Quality Ladder Model with Entry and Exit: Exploring the Equilibrium Correspondence Using the Homotopy Method," *Quantitative Marketing and Economics*, Vol. 10, 2012, pp. 197–229.

Congressional Budget Office, *Comparing the Compensation of Federal and Private-Sector Employees*, January 2012.

DMDC, *Civilian Personnel: Reports Related to Civilian Personnel Statistics*, web page, undated. As of May 15, 2014:
https://www.dmdc.osd.mil/appj/dwp/reports.do?category=reports&subCat=civReports

Edwards, Chris, "Overpaid Federal Workers," Washington, D.C.: Cato Institute, August 2013. As of May 17, 2014:
http://www.downsizinggovernment.org/overpaid-federal-workers

Falk, Justin, "Comparing Benefits and Total Compensation Between Similar Federal and Private-Sector Workers," *The B.E. Journal of Economic Analysis and Policy*, Vol. 12, Issue 1, 2012, pp. 1–35.

Gates, Susan M., Elizabeth Roth, Sinduja Srinivasan, and Lindsay Daugherty, *Analysis of the Department of Defense Acquisition Workforce: Update to Methods and Results Through FY 2011*. Santa Monica, Calif.: RAND Corporation, RR-110-OSD, 2013. As of October 24, 2013:
http://www.rand.org/pubs/research_reports/RR110.html

Gates, Susan M., Edward G. Keating, Adria D. Jewell, Lindsay Daugherty, Bryan Tysinger, Albert A. Robbert, and Ralph Masi, *The Defense Acquisition Workforce: An Analysis of Personnel Trends Relevant to Policy, 1993–2006*, Santa Monica, Calif.: RAND Corporation, TR-572-OSD, 2008. As of October 24, 2013:
http://www.rand.org/pubs/technical_reports/TR572.html

Gotz, Glenn, "Comment on 'The Dynamics of Job Separation: The Case of Federal Employees'," *Journal of Applied Econometrics*, Vol. 5, No. 3, 1990, pp. 1–35.

Gotz, Glenn A., and John McCall, *A Dynamic Retention Model for Air Force Officers: Theory and Estimates*, Santa Monica, Calif.: RAND Corporation, R-3028-AF, 1984. As of May 17, 2014:
http://www.rand.org/pubs/reports/R3028.html

Hotz, V. Joseph, and Robert Miller, "Conditional Choice Probabilities and the Estimation of Dynamic Models," *The Review of Economic Studies*, Vol. 60, No. 3, July 1993, pp. 497–529.

Kaiser Family Foundation, *2013 Employer Health Benefits Survey*, August 20, 2013. As of June 27, 2014:
http://kff.org/private-insurance/report/2013-employer-health-benefits/

Katz, Eric, "Democrats: Leave Federal Employees Alone," *Government Executive*, October 23, 2013. As of May 17, 2014:
http://www.govexec.com/pay-benefits/2013/10/democrats-leave-federal-employees-alone/72515/?oref=govexec_today_nl

Keane, Michael P., and Kenneth I. Wolpin, "The Career Decisions of Young Men," *Journal of Political Economy*, Vol. 105, No. 3, 1997, pp. 473–522.

Mattock, Michael G., and Jeremy Arkes, *The Dynamic Retention Model for Air Force Officers: New Estimates and Policy Simulations of the Aviator Continuation Pay Program*, Santa Monica, Calif.: RAND Corporation, TR-470-AF, 2007. As of May 17, 2014:
http://www.rand.org/pubs/technical_reports/TR470.html

Mattock, Michael G., Beth J. Asch, James Hosek, Christopher Whaley, and Christina Panis, *Toward Improved Management of Officer Retention: A New Capability for Assessing Policy Options*, Santa Monica, Calif.: RAND, TR-1260-OSD, 2014.

Mattock, Michael G., James Hosek, and Beth J. Asch, *Reserve Participation and Cost Under a New Approach to Reserve Compensation*, Santa Monica, Calif.: RAND Corporation, MG-1153-OSD, 2012. As of May 17, 2014:
http://www.rand.org/pubs/monographs/MG1153.html

Peters, Katherine McIntire, "The Drawdown Drags On," *Government Executive*, March 1, 1996. As of May 15, 2014:
http://www.govexec.com/magazine/1996/03/the-drawdown-drags-on/203/

Purcell, Patrick, *Federal Employees: Pay and Pension Increases Since 1969*, Washington, D.C.: Congressional Research Service, 94-971, updated January 8, 2008. As of May 16, 2014:
http://benefitslink.com/articles/guests/crs-94-971-2008.pdf

———, *Federal Employees: Pay and Pension Increases Since 1969*, Washington, D.C.: Congressional Research Service, 7-5700, January 20, 2010.

Rust, John, "Structural Estimation of Markov Decision Processes," in *Handbook of Econometrics*, Robert Engle and Daniel McFadden (eds.), Vol. IV, Elsevier Science B.V., 1994, pp. 3082–3143.

U.S. Bureau of Labor Statistics, *Highlights of Women's Earnings in 2012*, BLS Report 1045, Washington, D.C., October 2013. As of May 15, 2014:
http://www.bls.gov/cps/cpswom2012.pdf

U.S. Census Bureau, *Current Lists of Metropolitan and Micropolitan Statistical Areas and Delineations*, Washington, D.C., revised 2013. As of May 16, 2014:
http://www.census.gov/population/metro/data/metrodef.html

U.S. Department of Defense, *National Defense Budget Estimates for FY 2014*, Office of the Under Secretary of Defense (Comptroller), May 2013. As of May 17, 2014:
http://comptroller.defense.gov/Portals/45/Documents/defbudget/fy2014/FY14_Green_Book.pdf

U.S. Office of Personnel Management, *Healthcare: Eligibility*, web page, undated a. As of May 15, 2014:
http://www.opm.gov/healthcare-insurance/healthcare/eligibility/

———, *Pay & Leave: Salaries & Wages*, Washington, D.C., web page, undated b. As of July 1, 2014:
http://archive.opm.gov/oca/10tables/index.asp

———, *U.S. Bureau of Labor Statistics Data for Setting General Schedule Pay*, web page, undated c. As of May 16, 2014:
http://www.opm.gov/policy-data-oversight/pay-leave/salaries-wages/

van der Klaauw, Wilbert, "On the Use of Expectations Data in Estimating Structural Dynamic Choice Models," *Journal of Labor Economics*, Vol. 30, No. 3, July 2012, pp. 521–554.

van der Klaauw, Wilbert, and Kenneth Wolpin, "Social Security and the Retirement and Savings Behavior of Low-Income Households," *Journal of Econometrics*, Vol. 145, 2008, pp. 21–42.